Theory and Applications
of Natural Language Processing

Series Editors:
Graeme Hirst (Textbooks)
Eduard Hovy (Edited volumes)
Mark Johnson (Monographs)

Aims and Scope

The field of Natural Language Processing (NLP) has expanded explosively over the past decade: growing bodies of available data, novel fields of applications, emerging areas and new connections to neighboring fields have all led to increasing output and to diversification of research.

"Theory and Applications of Natural Language Processing" is a series of volumes dedicated to selected topics in NLP and Language Technology. It focuses on the most recent advances in all areas of the computational modeling and processing of speech and text across languages and domains. Due to the rapid pace of development, the diversity of approaches and application scenarios are scattered in an ever-growing mass of conference proceedings, making entry into the field difficult for both students and potential users. Volumes in the series facilitate this first step and can be used as a teaching aid, advanced-level information resource or a point of reference.

The series encourages the submission of research monographs, contributed volumes and surveys, lecture notes and textbooks covering research frontiers on all relevant topics, offering a platform for the rapid publication of cutting-edge research as well as for comprehensive monographs that cover the full range of research on specific problem areas.

The topics include applications of NLP techniques to gain insights into the use and functioning of language, as well as the use of language technology in applications that enable communication, knowledge management and discovery such as natural language generation, information retrieval, question-answering, machine translation, localization and related fields.

The books are available in printed and electronic (e-book) form:

* Downloadable on your PC, e-reader or iPad
* Enhanced by Electronic Supplementary Material, such as algorithms, demonstrations, software, images and videos
* Available online within an extensive network of academic and corporate R&D libraries worldwide
* Never out of print thanks to innovative print-on-demand services
* Competitively priced print editions for eBook customers thanks to MyCopy service http://www.springer.com/librarians/e-content/mycopy

For further volumes:
http://www.springer.com/series/8899

David Chiang

Grammars for Language and Genes

Theoretical and Empirical Investigations

Foreword by Aravind K. Joshi

 Springer

David Chiang
USC Information Sciences Institute
4676 Admiralty Way, Suite 1001
Marina del Rey, CA 90292
USA
chiang@isi.edu

Foreword by
Aravind K. Joshi
Department of Computer and Information Science
and
Institute for Research in Cognitive Science
University of Pennsylvania
Philadelphia, PA 19104
USA
joshi@seas.upenn.edu

ISSN 2192-032X   e-ISSN 2192-0338
ISBN 978-3-642-27080-2   ISBN 978-3-642-20444-9 (eBook)
DOI 10.1007/978-3-642-20444-9
Springer Heidelberg Dordrecht London New York

Springer is part of Springer Science+Business Media (www.springer.com)

JMJ

Foreword

It is indeed a great pleasure to write a few comments on this fascinating and inspiring book: **Grammars for Language and Genes: Theoretical and Empirical Investigations,** by David Chiang. First, I would like to acknowledge my good fortune in being able to work with David during his stay at the University of Pennsylvania. Each one of our meetings was a joyful event, informing and learning from each other, to our mutual benefit.

Chiang's work began with the study of the strong generative capacity of grammars, i.e., their capacity to represent structural descriptions. It is this aspect that is truly important for the study of formal grammars from the perspective of linguistics as well as computational linguistics. However, surprisingly, there is not much work done on issues concerning strong generative capacity (SGC). This is because it is not easy to formulate concepts of SGC that are formal enough and also linguistically meaningful. Building on notions of local interpretation functions, Chiang has given insightful accounts of how SGC should be characterized. He has then applied these ideas to a detailed study of characterizing SGC for a variety of formalisms including tree-adjoining grammars, their variants, and also several other formalisms. Further, building on some notions of extracting more SGC without increasing the weak generative capacity, Chiang has obtained some essential results connecting representations and interpretations. I am confident that much of this work will, in time, become the foundation on which to build further work on the formal characterizations of structural descriptions and interpretations and their eventual use in natural language processing (NLP).

The notion of squeezing more SGC without increasing the weak generative capacity plays a very significant role in the work described in the chapters on statistical parsing and machine translation. These investigations have been carried out in the general framework of tree-adjoining grammar (TAG) and some of its variants. I am sure researchers at large in statistical parsing and machine translation will be inspired by this work and will explore its implications for other classes of formal grammars, thus providing some unity in the very extensive work going on in these areas.

By a remarkable coincidence, just as Chiang was engaged in the activities described above, he also became a member of the group which began to explore the role of formal grammars in characterizing biomolecular structures, such as DNA/RNA and proteins, for example. This part of Chiang's book is a delightful treat for those who want to get a quick but thorough introduction to biomolecular structures and how to model a variety of these structures, keeping both the formal and computational aspects in mind at all times.

In summary, Chiang's work on grammars, which is based on solid mathematical foundations combined with a clear understanding of the domains that are being modeled, will lead to both a deeper theoretical understanding as well as usable computational models. I strongly recommend this book to all those who have already embarked on such activities but, more importantly, to those who would like to be involved in these exciting directions of research.

Aravind K. Joshi

Acknowledgements

This book was originally my PhD dissertation at the University of Pennsylvania. I would like to thank my adviser Aravind Joshi for his guidance during and after graduate school, and Ken Dill at UCSF who was an unofficial co-advisor for Chapters 5 and 6. My dissertation committee, Fernando Pereira, Jean Gallier, Sampath Kannan, and Giorgio Satta, provided a range of insightful perspectives; Giorgio Satta's careful reading was particularly helpful. Other faculty at Penn also gave me much valuable advice: Mark Steedman, Mitch Marcus, Martha Palmer, and David Searls.

Among the graduate students and postdocs I interacted with at Penn, I am particularly indebted to Dan Bikel, Mark Dras, and William Schuler. My collaborations with them and countless discussions taught me half of what I know about computational linguistics. The other half no doubt comes from the rest, including Mike Collins, Hoa Trang Dang, Dan Gildea, Julia Hockenmaier, Seth Kulick, Alexandra Kinyon, Tom Morton, Anoop Sarkar, and Fei Xia.

The revision of the thesis for publication incorporated material from a tutorial on synchronous grammars given with Kevin Knight, which benefited from feedback from Liang Huang and Philip Resnik. During the revision, I received many important comments and corrections from Steve DeNeefe, Jonathan May, and the anonymous reviewer.

I thank my sister Lillian and my future wife Juliana for their loving support, and my parents, who have given me more than I could possibly thank them for.

My graduate research was supported in part by the following grants: ARO DAAG5591-0228, NSF SBR-89-20230, and NSF ITR EIA-02-05456, awarded to the University of Pennsylvania.

Contents

Chapter 1
Introduction

Then the three young men of the bodyguard, who kept guard over the person of the king, said to one another, Let each of us state what one thing is strongest; and to the one whose statement seems wisest, King Darius will give rich gifts and great honors of victory... The first wrote, Wine is strongest. The second wrote, The king is strongest. The third wrote, Women are strongest, but above all things truth is victor.

1 Esdras 3.4–5, 10–12

One should realize... that if we consider these four, namely wine, the king, woman and truth, in themselves they are not comparable because they do not belong to the same genus. Nevertheless, if they are considered in relation to some effect, they coincide in one aspect, and so can be compared with each other.

St. Thomas Aquinas, *Quaestiones quodlibetales*, XII, q. 14, a. 1

1.1 The Problem of Strong Generative Capacity

Formal grammars, first developed as specifications of linguistic theories and programming languages, have found a rich variety of applications in computer science, especially in natural language processing and, more recently, biological sequence analysis. Grammars can be expressed in a variety of competing grammar formalisms, and the question naturally arises: What makes one grammar formalism better than another?

Formal language theory has traditionally given two ways of answering this question, namely, *weak generative capacity* (WGC) and *strong generative capacity* (SGC). The WGC of a grammar is the set of *strings* it generates, and its SGC is the set of *structural descriptions* it assigns to them. If we think of a grammar formalism as the set of all the grammars it can express, then the WGC (or SGC) of a formalism is the set of the WGCs (or SGCs) of its grammars, and we say that one formalism has greater WGC (or SGC) than another if its WGC (or SGC) is a superset of the other's.

Occasionally, one finds the term "strong generative capacity" misapplied to the set of phrase-structure trees a grammar generates (which we will refer to as its *tree generative capacity*). But a structural description may be any kind of structure a grammar might assign to a string: a phrase-structure tree, a dependency structure, an f-structure, a derivation tree, or a proof tree.

At the time that Chomsky introduced the terminology of weak and strong generative capacity, he observed that SGC is "by far the more interesting notion" [42, p. 297], but WGC is "the only area in which substantial results of a mathematical character have been achieved" [41, pp. 325–326]. The reason SGC is more interesting is that it is via structural descriptions that the grammar interfaces with higher-level modules (e.g., semantics). But the paucity of results having to do with SGC is due, at least in part, to the difficulty of defining what it means for two structural descriptions to be equivalent, especially when they come from different formalisms.

Nearly fifty years later, Chomsky's observation holds true. Formal language theory has produced many significant mathematical results, but continues to focus on WGC rather than SGC. Indeed, as more grammar formalisms are introduced, the more difficult it becomes to compare their structural descriptions and therefore their SGC. The problem of SGC has been addressed occasionally in the context of formal linguistics [76, 77, 89], but hardly at all in the context of computational applications, where the notion of SGC is no less relevant. Because results having to do with SGC are still lacking, the use of new grammar formalisms in these application areas is too often justified by intuition or examples or not at all.

1.2 Wine, King, Woman, and Truth

A similar problem was apparently faced by the three young men of the bodyguard of King Darius of Persia, the protagonists of the (literally apocryphal) story quoted at the beginning of this chapter. These three men argued for four different answers to the question, "What one thing is strongest?". Wine inebriates the greatest and least of people equally, the king commands everyone and they obey, and so on. Although the fourth answer, truth, prevailed in the end, St. Thomas Aquinas, commenting on this passage, maintains that each of the four arguments is valid in its own way. Like structural descriptions assigned by different grammar formalisms, these four things "are not comparable because they do not belong to the same genus."

Nevertheless, they can, Aquinas continues, be compared according to their effects in various domains. Wine has the strongest physical effect, woman has the strongest emotional effect, the king has the strongest effect on the practical intellect, and truth has the strongest effect on the speculative intellect. Thus, each domain has a different strongest thing. Likewise, even if we can't compare grammar formalisms directly, can we compare them according to their effects in some particular domain? This is the approach proposed by Miller [89]: to map structural descriptions generated by different formalisms into common *interpretation domains* where they can be compared. Thus, there is no single notion of SGC, but as many notions of SGC as there are interpretation domains. Here, we combine this approach with the formal approach of Joshi and collaborators, which we illustrate with an example.

1.3 The Case of Dutch Cross-Serial Dependencies

The controversy over the complexity of Dutch is a classic illustration of the kinds of issues involved in testing the adequacy of a grammar formalism, which will guide us as we develop the theoretical framework for our own comparisons of grammar formalisms. One early argument against the adequacy of context-free grammar (CFG) for natural language was put forth by Huybregts [61]. He argued that Dutch sentences exhibiting *cross-serial dependencies* like the following:

(1.1) dat Jan Piet de kinderen zag helpen zwemmen
      that Jan Piet the children saw help   swim
      that Jan saw Piet help the children swim

(where the first NP is the subject of the first verb, the second of the second, and so on) show that Dutch is like the copy language $\{ww\}$, which is non-context-free. Pullum and Gazdar [101] correctly replied that the sequence of verbs is not a copy of the sequence of nouns; the two sequences only had to be the same length. Therefore Dutch, considered as a set of strings, cannot be shown in any formal way to be reducible to the copy language.

Bresnan et al. [22] argued using traditional constituency arguments that the phrase-structure trees of sentences like (1.1) had to have a certain form, and then proved that CFG cannot generate such tree sets, concluding that CFG does not have enough "strong generative capacity" (in our terminology, tree generative capacity) to capture this construction. However, because it relied on theory-internal assumptions to determine the desired trees, this argument was not totally compelling.

Finally, Huybregts [62] and Shieber [122] independently observed that Swiss German allows a cross-serial word order as Dutch does but also has verbs which mark their objects with different cases.

(1.2) das mer d'chind   em Hans   es huus   lönd hälfe aastriiche
      that we the children-ACC Hans-DAT the house-ACC let   help paint
      that we let the children help Hans paint the house

(1.3) *das mer d'chind   de Hans   es huus   lönd hälfe aastriiche
      that we the children-ACC Hans-ACC the house-ACC let   help paint
      that we let the children help Hans paint the house

Therefore Swiss German can be reduced via homomorphisms and intersections to the copy language $\{ww \mid w \in a^*b^*\}$, which proves that Swiss German is not a context-free language. But this still does not settle the question for Dutch.

The most satisfying answer to the question of Dutch comes from the literature on tree-adjoining grammars (TAG) and related formalisms. Just as CFGs generate strings by rewriting symbols as strings, *tree-adjoining grammars* [69, 70] generate trees by rewriting nodes as trees.

For example, consider the TAG of Figure 1.1. The tree α is called an *initial tree*, roughly analogous to the start symbol in CFG. The trees β_1 and β_2 are called

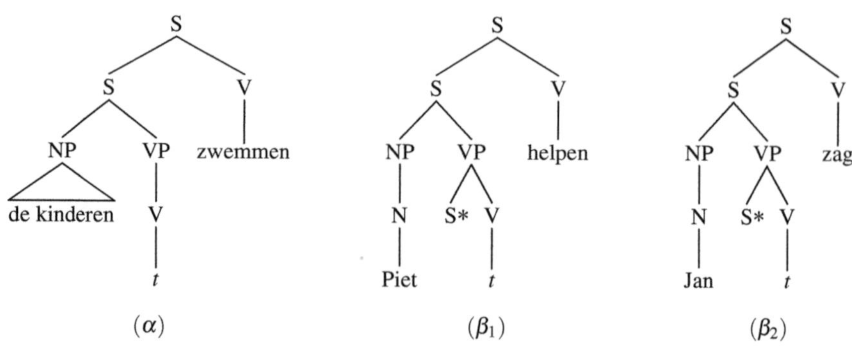

Fig. 1.1 TAG for Dutch cross serial dependencies in sentence (1.1).

auxiliary trees, analogous to productions in CFG. They differ from initial trees in that they have exactly one frontier node marked with an asterisk (∗); this node is called the *foot node* and always has the same label as the root node. The path from the root node to the foot node of an auxiliary tree is called its *spine*.

The basic rewriting operation is called *adjunction,* in which a node is rewritten with the spine of an auxiliary tree β along with all its branches. The rewritten node and the root/foot of β must have the same label. For example, Figure 1.2 shows the result of rewriting the lower S node of α with β_1.

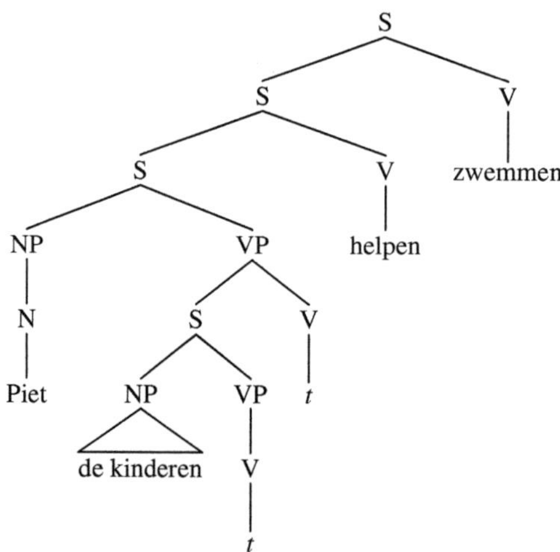

Fig. 1.2 Example of adjunction. The tree β_1 has been adjoined at the lower S node of α.

Joshi [67] showed that a tree-adjoining grammar with *links* (see Figure 1.3; this particular analysis is due to Kroch and Santorini [75]) could generate examples

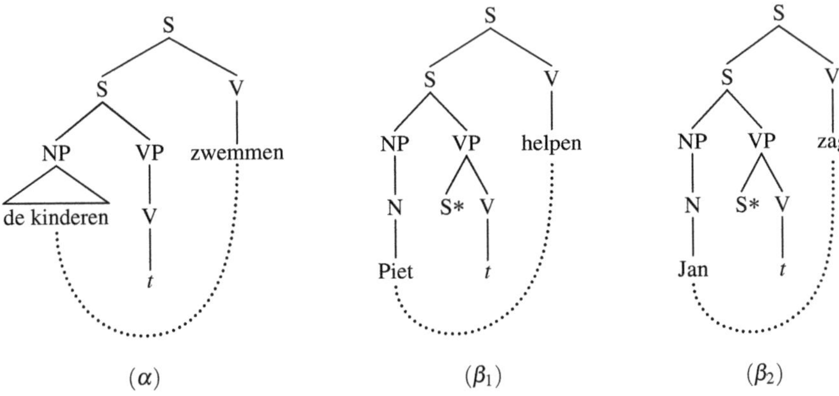

Fig. 1.3 TAG with links for Dutch cross-serial dependencies in sentence (1.1).

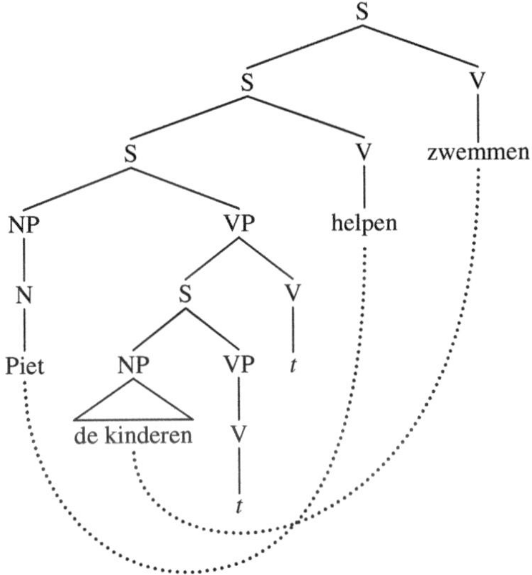

Fig. 1.4 First step in derivation of cross-serial dependencies. The tree β_1 has been adjoined at the lower S node of α.

like (1.1) with the dependencies (which are much less controversial than the phrase structure) explicitly marked:

(1.4)   dat Jan Piet de kinderen zag helpen zwemmen

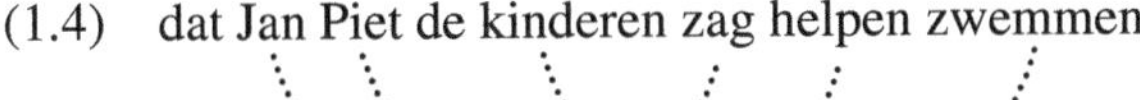

Figures 1.4 and 1.5 show the derivation of the sentence, with the cross-serial dependencies as desired. This is possible with a TAG but not possible for any CFG. There

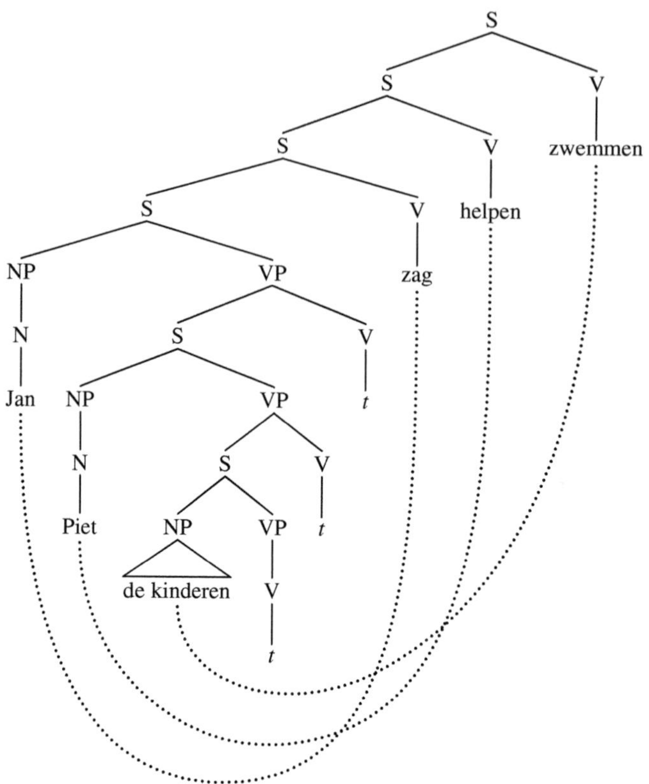

Fig. 1.5 Second step in derivation of cross-serial dependencies. The tree β_2 has been adjoined at the third S node of Figure 1.4.

are two key assumptions at work in this argument. First, grammars are not measured according to their WGC but according to their ability to generate strings with subject-verb dependencies explicitly marked with links. Second, CFGs and TAGs can only generate links between terminal symbols in the same elementary structure.

Becker et al. [11] developed these two assumptions into the notion of *derivational generative capacity* as an alternative to weak generative capacity and tree generative capacity.

Definition 1. A *linked string* is a pair $\langle w, \sim \rangle$, where w is a string and $\sim$ is a symmetric binary relation between string positions of w.

Definition 2. We say that a grammar G generates a linked string $\langle w, \sim \rangle$ if G generates w such that $i \sim j$ if and only if the ith and jth symbol of w are generated in the same derivation step.

Definition 3. The *derivational generative capacity* (DGC) of G is the set of all linked strings generated by G.

We notate linked strings either using arcs as above, or sometimes boxed numbers ($\boxed{1},\boxed{2},\ldots$) if the linking relation is transitive. In this terminology, then, we would say that CFG lacks the DGC to capture cross-serial dependencies in Dutch.

The merit of this approach is that it eschews notions of generative capacity which are inappropriate (WGC, tree generative capacity) or vague (SGC) in favor of a notion (DGC) which allows a rigorous result to be proven from minimal assumptions (namely, what the correct dependencies are and how dependencies must be represented in a CFG). Returning to Miller's idea that SGC should always be measured relative to some interpretation domain, we can view DGC, then, as one of many possible notions of SGC (i.e., SGC with respect to the domain of linked strings), the one that is best suited to the application at hand.

1.4 Overview

Theoretical framework

In Chapter 2 we set up the basic framework in which we will carry out our comparisons of grammar formalisms. We show how to generalize DGC to other interpretation domains using the concept of *local interpretation functions*. This allows us to rigorously classify a wide range of grammar formalisms according to their power in various interpretation domains. Different interpretation domains classify formalisms differently: some will be coarser-grained, some will be finer-grained; it can even happen that one formalism is more powerful than another in one domain, but less powerful in another domain. We are especially interested in situations where we can "squeeze" SGC out of a formalism—that is, to increase the SGC of a formalism while preserving its computational properties [68]. We try to capture this intuition using the notion of a *cover* [94]: a situation when one grammar is parsed using another grammar (the cover grammar) and therefore inherits its computational properties.

We will explore three areas of application: statistical natural language parsing, natural language translation, and modeling of biological macromolecules. For each area, we will define an appropriate interpretation domain, compare various grammar formalisms with respect to that domain, and then explore the implications of these results for practical applications.

Statistical parsing

We first consider, in Chapter 3, the task of statistical parsing: computing the most likely structure (standardly, the most likely phrase-structure tree) of a given string. We discuss how to define statistical models based on a large class of grammar formalisms. We measure statistical-modeling power using SGC with respect to the domain of weighted parse structures, which turns out to classify formalisms rather coarsely: if a grammar G can be covered by (say) a CFG, then weights can also be assigned to the cover grammar to make it strongly equivalent to G with respect to weighted parses. In other words, we do not expect in general that formalisms which squeeze SGC out of CFG will provide any more statistical parsing models than weighted CFG does (although a more precise treatment of the question below leaves some room for exceptions).

But if the statistical-modeling power of these squeezed formalisms is already accessible within PCFG, then we can investigate how that power may already be utilized by existing PCFG models like those of Charniak [25] or Collins [44], which represented a breakthrough in statistical parsing. It turns out that the style of CFGs, called *lexicalized CFGs*, that these models use is very similar to the cover CFGs from the above result. Thus, applying the construction in reverse to a lexicalized PCFG model yields a reinterpretation of lexicalized PCFG as a special kind of probabilistic TAG.

But TAG structural descriptions contain more information than the phrase-structure trees found in typical training corpora like the Penn Treebank [87]. Under this interpretation, then, it becomes more clear that the purpose of the head-propagation rules used by lexicalized PCFG models is not simply to rearrange information in the training data, but to *reconstruct* information missing from the training data; and this information is not lexical, but *structural*.

We then describe our implementatation of this interpretation in a parsing model based on probabilistic tree-insertion grammar with sister-adjunction (TIG-SA). We reinterpret the head/argument rules from Magerman's SPATTER parser [86] and Collins' parser [44] as a heuristic for reconstructing full structural descriptions from partial ones; we also explore a method related to the approach of Hwa [63] which uses Expectation-Maximization to directly estimate the model defined over full structural descriptions on the partial structural descriptions in the training data. We present experimental results for both of these techniques, training our probabilistic TIG-SA model from the Penn Treebank (English) and the Penn Chinese Treebank. We find that our probabilistic TIG-SA model performs at roughly the same level as lexicalized PCFG models and explore some new directions of research that such a model opens up.

Machine translation

The next application we discuss is translation (Chapter 4). In contrast to the classification above, this notion of SGC classifies formalisms quite finely. A number of formalisms that have been proposed in the literature are weakly equivalent to CFG, but differ in their translation power, that is, their SGC with respect to the domain of string or tree pairs. To these we add a new formalism, synchronous regular-form

TAG [49, 39, 33]. We also take a look at synchronous TAG and extensions of synchronous TAG.

Machine translation using synchronous CFGs and synchronous tree-substitution grammars is now a fairly well-explored area. We provide a survey of recent research from a formal-grammars perspective. An area that continues to be challenging is tree-to-tree translation, that is, the modeling of two parallel syntactic structures in different languages. We discuss a few cases where the additional translation power of richer formalisms like synchronous regular-form TAG might prove useful. We argue that because translation power classifies formalisms so much more finely than statistical-modeling power, it is more important in translation research to target the right formalism.

Biological sequence analysis

Finally, in Chapter 5 we explore the use of formal grammars for biological sequence analysis. Proteins and RNAs are folded from molecules which are chains of building blocks (amino acids and nucleotides) assembled in a sequence specified by genes. The task of biological sequence analysis is to relate genetic sequences to the folded structures they encode. It was Searls [120] who first observed the similarity between biological sequence analysis and natural-language-syntactic analysis and proposed that the same techniques could be applied to both problems. We give a unified treatment of previous applications of formal grammars to this problem, highlighting in particular their shared assumption that grammatical locality corresponds to physical locality. This observation implies that the relevant notion of SGC for this problem is that of linked strings. Searls' original work was on CFG; we explore some ways of employing formalisms with greater SGC than CFG to model more complex structures: α-helices, β-sheets, kissing hairpins, and pseudoknots.

Most grammatical approaches to biological sequence analysis rank structures using weights—usually probabilities or energies. We describe a more sophisticated use of weights, drawing on a model due to Chen and Dill [28, 29] which tries not only to predict structures of chain molecules but to give a full description of their statistical-mechanical properties. Their model is not explicitly grammatical, but we show that it can be more cleanly viewed as a weighted CFG.

In Chapter 6, we explore a family of approaches based on the technique of *intersection*—analyzing a string with two or more grammars and composing their structural descriptions. Intersection is not used much for natural language, probably because hierarchical structural descriptions do not compose easily, but is more promising for biological sequence analysis, because there is a well-defined way of composing structures. We show how our CFG version of Chen and Dill's model can be intersected with a finite-state automaton for α-helices, easily yielding a novel model for bundles of α-helices. We also discuss how simple literal movement grammars [57] (similar to range concatenation grammars [20]) might use their built-in intersection operation to efficiently model protein β-sheets.

Chapter 2
Foundation

In this chapter we define the framework that will be used for the investigations in subsequent chapters. Our framework combines and extends the approaches of Miller [89] and of Joshi and collaborators. We take the position, following Miller, that SGC should always be tested with reference to a particular *interpretation domain*, and, following Becker et al. [11], we define these interpretations on grammar formalisms with well-defined domains of *locality*, permitting rigorous formal comparisons. Moreover, we extend these ideas to a still larger class of formalisms, simple literal movement grammars [57].

2.1 Strong Generative Capacity, Relativized

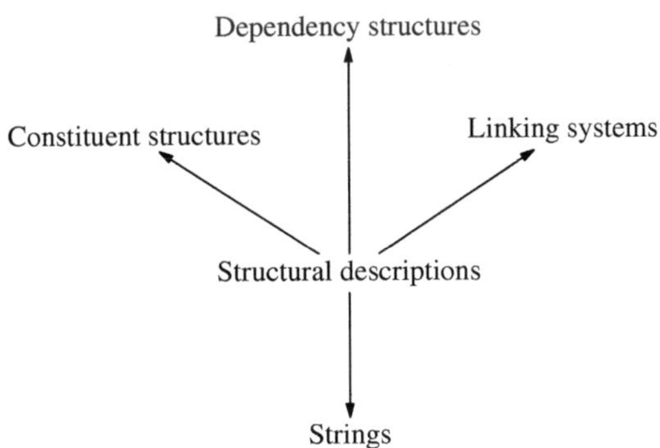

Fig. 2.1 Structural descriptions, strings, and interpretations under Miller's framework.

Miller [89] deals with the elusiveness of SGC by defining it not as the set of structural descriptions of a grammar, but the set of *interpretations* of its structural descriptions in a particular *interpretation domain*, for example, constituency or dependency (see Figure 2.1). Thus SGC must always be considered with respect to an interpretation domain. This creates many notions of SGC, but ensures that the SGC of two formalisms can be meaningfully compared—provided they both have interpretation functions in a common domain.

A strength of Miller's approach is that it tries to preserve the generality of the formalisms that can be dealt with, placing no restrictions on structural descriptions or their interpretation functions except that interpretation functions are not supposed to add information to structural descriptions that is not specified by the formalism. But this generality makes it hard to make definite statements about what a formalism is capable of.

The approach of Joshi and of Becker et al., which can be seen as a special case of Miller's, provides more traction. The key to this approach is the concept of *locality*. DGC, as we saw in the previous chapter, assumes that structural descriptions can be decomposed into elementary structures, and that the interpretation of a structural description can be defined locally on each elementary structure: only symbols in the same elementary structure can be linked. This constrains the interpretation functions for a large class of formalisms in such a way that it can be rigorously shown that, for example, Dutch cross-serial dependencies are possible for TAG but impossible for CFG.

We generalize the TAG approach to allow for many interpretation domains, retaining the key idea of specifying *local interpretation functions* on elementary structures. In the following section we define local interpretation functions more precisely.

2.2 Simple Literal Movement Grammars

There are many grammar formalisms for which interpretation functions can be decomposed into local interpretation functions on elementary structures; all the formalisms we will consider are subclasses of Groenink's *simple literal movement grammar*, or sLMG [57]. Simple LMGs are a special type of Post system [99, 42]. They resemble Prolog programs where the variables and constants range over strings and can be concatenated. They are equivalent to range concatenation grammar or RCGs [20], except that RCG variables stand for ranges of positions of a fixed input string, whereas sLMG variables stand for strings.

As a nonlinguistic example, here is a grammar which accepts all and only the Fibonacci numbers, written in unary notation:

$$\text{Fib}(y) :- \text{Fib2}(x,y) \tag{2.1}$$

$$\text{Fib2}(y,xy') :- \text{Fib2}(x,y), \text{Eq}(y,y') \tag{2.2}$$

$$\text{Fib2}(1,1) \tag{2.3}$$

$$\text{Eq}(1x,1x') :- \text{Eq}(x,x') \tag{2.4}$$

$$\text{Eq}(\varepsilon,\varepsilon) \tag{2.5}$$

The predicate Eq holds of two numbers (written in unary notation) just in case they are equal. Clause (2.5) is an axiom that says that $0 = 0$; clause (2.4) says that $x+1 = x'+1$ if $x = x'$. The predicate Fib2 holds of two numbers (written in unary notation) just in case they are consecutive Fibonacci numbers. Clause (2.3) is an axiom that says that 1 and 1 are consecutive Fibonacci numbers. Clause (2.2) says that y and $x+y'$ are consecutive Fibonacci numbers if x and y are and $y = y'$. Finally, clause (2.1) says that y is a Fibonacci number if x and y are consecutive Fibonacci numbers. Then the proof that 5 (11111 in unary) is a Fibonacci number proceeds as follows (leaving out the subderivations for Eq):

$$
\begin{array}{ll}
\text{Fib2}(1,1) & \text{from (2.3)} \\
\text{Fib2}(1,11) & \text{from (2.2) with } x = 1, y = y' = 1 \\
\text{Fib2}(11,111) & \text{from (2.2) with } x = 1, y = y' = 11 \\
\text{Fib2}(111,11111) & \text{from (2.2) with } x = 11, y = y' = 111 \\
\text{Fib}(11111) & \text{from (2.1) with } x = 111, y = 11111
\end{array}
$$

The TAG of Figure 1.1 would be encoded as the following sLMG:

$$
\begin{array}{lr}
\text{S}(x \text{ de kinderen } y \text{ zwemmen}) :- \text{S2}(x,y) & (\alpha) \\
\text{S2}(x \text{ Piet}, y \text{ helpen}) :- \text{S2}(x,y) & (\beta_1) \\
\text{S2}(x \text{ Jan}, y \text{ zag}) :- \text{S2}(x,y) & (\beta_2) \\
\text{S2}(\varepsilon,\varepsilon) & (\varepsilon)
\end{array}
$$

The unary predicate S holds of the string yields of derived initial trees rooted by S. The binary predicate S2 holds of the string yields (to the left and to the right of the foot) of derived auxiliary trees rooted by S. The derivation of (1.1) proceeds as follows:

$$
\begin{array}{ll}
\text{S2}(\varepsilon,\varepsilon) & \text{from } (\varepsilon) \\
\text{S2}(\text{Jan}, \text{zag}) & \text{from } (\beta_2) \\
\text{S2}(\text{Jan Piet}, \text{zag helpen}) & \text{from } (\beta_1) \\
\text{S}(\text{Jan Piet de kinderen zag helpen zwemmen}) & \text{from } (\alpha)
\end{array}
$$

We now give a more formal definition.

Definition 4. A *simple LMG* is a tuple $\langle T,N,V,S,A,P \rangle$, where:

- T is a finite set of *terminal symbols* and N is a finite set of *nonterminal symbols*
- V is a set of *variables*
- $S \in N$ is called the *start symbol*

- A is a set of axioms of the form

$$X(\alpha_1, \ldots, \alpha_m)$$

where $X \in N$ and $\alpha_j \in T^*$
- P is a set of productions of the form

$$X(\alpha_1, \ldots, \alpha_m) :- Y_1(\beta_{11}, \ldots, \beta_{1m_1}), \ldots, Y_n(\beta_{n1}, \ldots, \beta_{nm_n})$$

("the α_i are an X if the β_{1i} are a Y_1 and the β_{2i} are a Y_2, etc.") where

- $X, Y_i \in N$
- $\alpha_j \in (T \cup V)^*$
- $\beta_{ij} \in V$
- each variable in the production appears exactly once on the left-hand side and at least once on the right-hand side

Most sLMGs used in natural language processing belong to the class of *linear* sLMGs, which is equivalent to LCFRS, nonerasing multiple context-free grammar [121], local scattered context grammar [102], and simple RCG [20].

Definition 5. A *linear sLMG* is an sLMG in which for each production, each variable in the production appears exactly once on the right-hand side and exactly once on the left-hand side.

An sLMG defines a deductive proof system whose theorems are statements about relations between strings; the start predicate, which is always unary, holds of all and only the strings of the language defined by the grammar.

Definition 6. Let π be an sLMG production, $x_1, \ldots, x_n$ be all the distinct variables occurring in π, and $w_1, \ldots, w_n \in T^*$. Let π' be the result of substituting w_i for x_i for all $1 \leq i \leq n$; then we say that π' *instantiates* or is an *instantiation* of π.

Definition 7. If G is an sLMG, we say that G *derives* $X(\alpha_1, \ldots, \alpha_m)$ according to the following recursive definition:

- An axiom $X(\alpha_1, \ldots, \alpha_m)$ of G is derivable by G.
- $X(\alpha_1, \ldots, \alpha_m)$ is derivable by G if there is a production in G which can be instantiated as

$$X(\alpha_1, \ldots, \alpha_m) :- Y_1(\beta_{11}, \ldots, \beta_{1m_1}), \ldots, Y_n(\beta_{n1}, \ldots, \beta_{nm_n})$$

and $Y_i(\beta_{i1}, \ldots, \beta_{im_i})$ is derivable by G for $1 \leq i \leq n$.
- Nothing else is derivable by G.

Definition 8. The weak generative capacity $L(G)$ of G is the set $\{w \mid G \text{ derives } S(w)\}$.

We also sometimes represent the derivation process as a tree:

Definition 9. A *derivation* of an sLMG G is a tree over (names of) productions of G, defined recursively:

- If $\pi = X(\alpha_1,\ldots,\alpha_m)$ is an axiom of G, $\pi()$ is a derivation of $X(\alpha_1,\ldots,\alpha_m)$.
- If a production π of G can be instantiated as

$$X(\alpha_1,\ldots,\alpha_m) :- Y_1(\beta_{11},\ldots,\beta_{1m_1}),\ldots,Y_n(\beta_{n1},\ldots,\beta_{nm_n})$$

and τ_i is a derivation of $Y_i(\beta_{i1},\ldots,\beta_{im_i})$, $1 \le i \le n$, then $\pi(\tau_1,\ldots,\tau_n)$ is a derivation of $X(\alpha_1,\ldots,\alpha_m)$.

Let $\mathscr{D}(G)$ denote the set of derivations of G.

2.3 Interpretation Functions

Next, we describe how to define interpretations of sLMG derivations. We do this by attaching a local interpretation function to each clause of the grammar, in a manner similar to attribute grammars [73].

Definition 10. An n-ary *local interpretation function* in D is any function from D^n to D. A 0-ary local interpretation function on D just returns a constant member of D.

Definition 11. An *interpretation function* for an sLMG G in interpretation domain D is a function $f : \mathscr{D}(G) \to D$. An assignment of a local interpretation function f_π to each production $\pi \in P$ induces an interpretation function f for G:

- If π is an axiom, $f(\pi()) = f_\pi()$.
- If π is a production of the form

$$X(\alpha_1,\ldots,\alpha_m) :- Y_1(\beta_{11},\ldots,\beta_{1m_1}),\ldots,Y_n(\beta_{n1},\ldots,\beta_{nm_n}),$$

then $f(\pi(\tau_1,\ldots,\tau_n)) = f_\pi(f(\tau_1),\ldots,f(\tau_n))$.

For example, to extend our sLMG for Dutch to generate links as in Figure 1.3, we define local interpretation functions for each clause:

clause	local interpretation function
(α)	w_1 de kinderen w_2 zwemmen $\leftarrow\!\dashv \langle w_1, w_2 \rangle$
(β_1)	$\langle w_1$ Piet, w_2 helpen$\rangle \leftarrow\!\dashv \langle w_1, w_2 \rangle$
(β_2)	$\langle w_1$ Jan, w_2 zag$\rangle \leftarrow\!\dashv \langle w_1, w_2 \rangle$
(ε)	$\langle \varepsilon, \varepsilon \rangle$

where the notation $y \leftarrowtail x$ denotes a local interpretation function that maps x to y. The interpretation of our example derivation is the linked string (1.4).

Definition 12. An *interpretation function* Φ for a grammar formalism $\mathscr{F}$ in domain D assigns to each production allowed in $\mathscr{F}$ a set of possible local interpretation functions on D. This induces a mapping from each grammar $G \in \mathscr{F}$ to a set, $\Phi(G)$, of possible interpretation functions for G in D.

The reason Φ assigns a set of possible interpretation functions, rather than a single interpretation function, is that more than one interpretation function may be reasonable. For example, the word *de* in clause (α) could be linked to *kinderen*, or to *zwemmen*, or nothing.[1] This is a departure from Miller, whose interpretation functions map each derivation of $\mathscr{F}$ to a single interpretation. We do, however, want Φ to restrict the kinds of local interpretation functions that can be used. For example, in the interpretation domain of linked strings (as used here and in Chapter 5), we don't want a local interpretation function to change or delete any links in its arguments.

We now have the necessary framework for our definition of strong generative capacity.

Definition 13. The strong generative capacity $\Sigma_D(G, f)$ of G with respect to the interpretation domain D and interpretation function f is the set

$$\{f(\tau) \mid \tau \text{ is a derivation of } G\}$$

and the strong generative capacity of a formalism $\mathscr{F}$ with respect to D and interpretation function Φ is the set of sets

$$\{\Sigma_D(G, f) \mid G \in \mathscr{F} \text{ and } f \in \Phi(G)\}.$$

We define a trivial interpretation domain, that of strings, in which the interpretation of every derivation is just the string yield of that derivation:

Definition 14. The interpretation domain of strings has interpretations which are (tuples of) strings, and local interpretation functions defined as follows: for each axiom π of the form $X(\alpha_1, \ldots, \alpha_m)$, the local interpretation function must be $f_\pi() = \langle \alpha_1, \ldots, \alpha_m \rangle$, and for each production π of the form

$$X(\alpha_1, \ldots, \alpha_m) :- Y_1(\beta_{11}, \ldots, \beta_{1m_1}), \ldots, Y_n(\beta_{n1}, \ldots, \beta_{nm_n})$$

the local interpretation function must be

$$f_\pi(\langle w_{11}, \ldots, w_{1m_1} \rangle, \ldots, \langle w_{n1}, \ldots, w_{nm_n} \rangle) = \langle \alpha_1[w_{ij}/\beta_{ij}], \ldots, \alpha_m[w_{ij}/\beta_{ij}] \rangle$$

where $[w_{ij}/\beta_{ij}]$ means, "replace β_{ij} with w_{ij}." If the sLMG is nonlinear, this substitution may not always be consistent; in such cases, which are harmless, f_π is undefined.

---

[1] In practice, there may be situations where we want to assign multiple local interpretation functions to a production, that is, different local interpretation functions to multiple identical productions. For simplicity, we have not attempted to cover this possibility.

For the interpretation domain of trees, the interpretation functions will vary by formalism. For example, CFG derivations would simply be mapped to themselves, whereas TAG derivations would map to derived trees as in Figure 1.2. The only general statement we can make is that a derivation's interpretation in the domain of trees should yield the same string as its interpretation in the domain of strings.

Finally, we restate the definition of DGC in the new framework, slightly relaxed to allow unlinked symbols:

Definition 15. The interpretation domain of linked strings has interpretations which are (tuples of) linked strings (Definition 1), and whose local interpretation functions are the same as for the domain of strings except that terminal symbols in the α_i may be linked to each other (cf. Definition 2). The *derivational generative capacity* of a grammar or grammar formalism is its SGC with respect to the domain of linked strings.

2.4 Summary

The framework laid out in this chapter provides the control needed to prove results with genuine relevance to applications. Because SGC is defined as interpretations with reference to particular domains, we can test the relevant properties of a formalism; because interpretations are defined in terms of local interpretations, we can firmly characterize a formalism's SGC.

In subsequent chapters, we will use this framework to define interpretation domains suited to particular applications and then compare various grammar formalisms in those interpretation domains. The goal is to see whether more results like those of Joshi and collaborators can be obtained in these other areas of application, and what implications they have for those applications.

2.5 Additional Topics

This section contains some supplemental material which will be referred to in subsequent chapters, and which the impatient reader may safely skip on a first reading. Sections 2.5.1 and 2.5.2 contain definitions of various formalisms that are subclasses of sLMG, and Section 2.5.3 contains a brief treatment of sLMG parsing. Section 2.5.4 introduces the concept of cover grammars, which will be referred to mainly in Chapter 3.

2.5.1 Tree-Adjoining Grammars

We have already introduced tree-adjoining grammars in Section 1.3, but this section presents a more complete definition of TAGs and several variants which will be referred to throughout.

Definition 16. An *auxiliary tree* is a finite tree with a distinguished frontier node called its *foot node*, which we mark with the symbol $*$. The path from an auxiliary tree's root node to its foot node is called its *spine*.

Definition 17. A *tree-adjoining grammar* [69, 70] is a tuple $\langle T, N, I, A, S \rangle$, where

- T is a finite set of terminal symbols
- N is a finite set of nonterminal symbols, $N \cap T = \emptyset$
- $N' = N \times \{\varepsilon, \text{NA}, \text{OA}\}$ is the set of nonterminal symbols with *adjoining constraints*; unless otherwise indicated, equivalence is understood to be modulo adjoining constraints
- I is a finite set of *initial trees*, which are finite trees whose interior labels are drawn from N' and whose frontier labels are drawn from $N' \cup T$
- A is a finite set of auxiliary trees whose interior labels are drawn from N', whose frontier labels are drawn from $N' \cup T$, and whose root and foot nodes bear the same label
- $S \subseteq I$ is a set of initial trees which can begin a derivation

Definition 18. The result of *adjoining* an auxiliary tree β with root/foot label X at a node η with label X is the tree obtained as follows: detach the subtree rooted by η and call it γ_η, leaving behind a copy of η; attach β by merging its root node with (the copy of) η; reattach γ_η by merging its root node with the foot node of β.

Definition 19. The result of *substituting* an initial tree α with root label X at a frontier node η with label X is the tree obtained by merging the root node of α with η.

Definition 20. A *derived tree* (or *derived initial tree* or *derived auxiliary tree*) of G is obtained by taking an elementary tree γ in S (or I or A, respectively), and:

- substituting a derived initial tree at each of the non-foot frontier nonterminal nodes (called *substitution nodes*)
- adjoining a derived auxiliary tree at each of the nodes with adjoining constraint OA and zero or more of the nodes without adjoining constraint NA

Definition 21. The tree set or tree generative capacity of a TAG G is the set of all possible derived trees of G. The string set or weak generative capacity of G is the set of yields of derived trees of G.

The following three restrictions of TAG have been proposed to capture some of the additional descriptive power of TAG while remaining weakly context-free.

No adjunction: tree substitution grammars

Definition 22. A *tree-substitution grammar* or TSG [111] is a tree-adjoining grammar with no auxiliary trees.

As a historical note, TAG as originally defined only had adjunction; substitution was introduced later by Abeillé [2], and then adjunction was dropped by Schabes et al. [113] to form TSG, though it does not seem to have been called by that name until later [111].

No wrapping adjunction: tree-insertion grammars

Definition 23. A *left* (or *right*) *auxiliary tree* is an auxiliary tree in which no frontier node lies to the right (or left, respectively) of the foot node. (If we permit leaf nodes labeled with empty elements, these may lie anywhere.)

Definition 24. A *tree-insertion grammar* or TIG [115, 116], originally termed a "lexicalized context-free grammar," is a TAG in which all auxiliary trees are either left or right auxiliary trees, and adjunction is constrained so that:

- no left (right) auxiliary tree can be adjoined on any node that is on the spine of a right (left) auxiliary tree, and
- no adjunction is permitted on a node that is to the right (left) of the spine of a left (right) auxiliary tree.

Limited spine adjunction: regular form

In his original definition, the details of which we omit here, Rogers [107] defines a restriction on TAG adjunction, called *regular adjunction*, that can generate only regular path sets. He then identifies a subclass of TAGs, called TAGs in *regular form*, which have the property that every derived tree that can be derived using unrestricted adjunction could also have been derived using only regular adjunction. But since Rogers' recognition algorithm only performs regular adjunction, it cannot in general produce all possible derivations of a sentence and therefore cannot be used as a parser.

A more technical issue is that regular adjunction can occur at either the root or foot, which creates derivational ambiguity. Rogers' algorithm, however, cannot distinguish between the two. If we want the parser to compute derivations, one or the other should be disallowed. Following Schuler et al. [119], we prohibit adjunction at

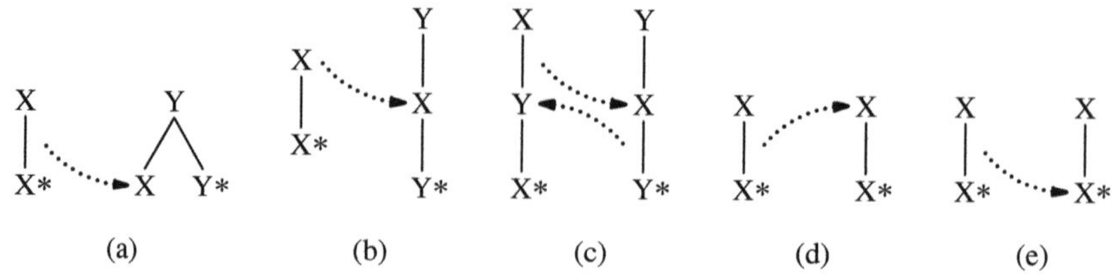

Fig. 2.2 Examples of adjunction in regular-form TAG. (a) Off-spine adjunction, allowed; (b) acyclic spine adjunction, allowed; (c) cyclic spine adjunction, not allowed; (d) root adjunction, not allowed; (e) foot adjunction, allowed.

the root.[2] This leads us to the following definition, which narrows Rogers' definition to eliminate both of the above problems:

Definition 25. We say that a TAG is in *regular form*, or an RF-TAG, if there exists some partial ordering $\preceq$ over nonterminal symbols such that if β is an auxiliary tree whose root and foot nodes are labeled X, and η is a node labeled Y on β's spine where adjunction is allowed, then $X \preceq Y$, and $X = Y$ only if η is a foot node.

Thus adjunction at nodes not lying along the spine and adjunction at the foot node are allowed freely; adjunction at nodes lying along the spine is allowed to a bounded depth, but adjunction at the root is not allowed at all (see Figure 2.2).

Multiple adjunction and sister adjunction

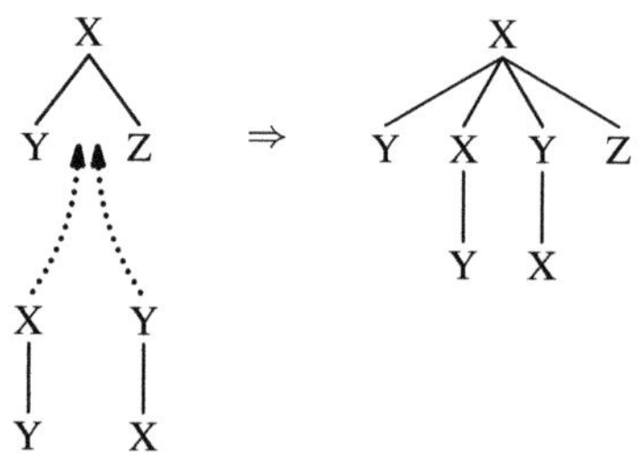

Fig. 2.3 Sister adjunction.

In standard definitions of TAG, only one auxiliary tree is allowed to adjoin at a single node. Schabes and Shieber [114] argue that this is unsatisfactory for certain

---

[2] If we had prohibited adjunction at the foot, as is more customary, and allowed adjunction at the root, then the resulting grammars would not be coverable by CFGs (see Section 2.5.4). It might be possible to relax the definition of a cover grammar to allow this, but we do not pursue this possibility here.

linguistic constructions, and so they propose an extended notion of derivation in which a distinction is made between *modifier auxiliary trees* and *predicative auxiliary trees*. Multiple modifier auxiliary trees may be adjoined at a single node, but only one predicative auxiliary tree may be adjoined at a single node.

We combine the idea of multiple adjunctions with an operation borrowed from d-tree substitution grammar [103] called *sister-adjunction*:

Definition 26. The result of *sister-adjoining* an initial tree α under a node η at position i is the tree obtained by

- if $i = 0$: adding α as the leftmost daughter of η;
- if $0 < i < n$, where n is the number of daughters of η: inserting α between the ith and $(i+1)$st daughter of η;
- if $i = n$: adding α as the rightmost daughter of η.

See Figure 2.3. As in Schabes and Shieber's extension, multiple trees may be sister-adjoined at the same position [31].

This extension does not add any weak generative power. However, a TAG extended in this way is no longer an sLMG, strictly speaking, because its derivation trees can have unbounded branching factors, whereas an sLMG's derivation trees only allow bounded branching.

2.5.2 Multicomponent Grammars

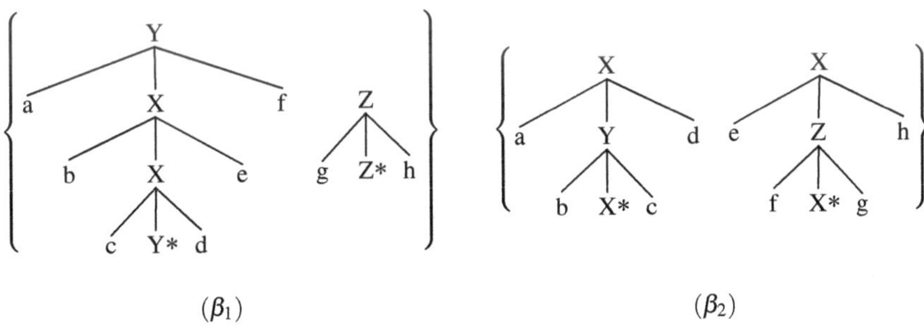

Fig. 2.4 Example multicomponent TAG elementary tree sets.

Multicomponent TAGs [129] are TAGs whose elementary structures are sets of elementary trees. The basic operation is the simultaneous adjunction or substitution of all the trees in a set. In *set-local* multicomponent TAG, all the trees must compose into the same elementary tree set; in *tree-local* multicomponent TAG, all the trees must compose into the same elementary tree. For example, Figure 2.4 shows some multicomponent TAG elementary tree sets. In a set-local multicomponent TAG, β_1 would be able to adjoin into β_2, by adjoining the first component into the first component and the second component into the second component. Moreover, β_2 would be able to adjoin into β_1, by adjoining both components into the two X nodes of the first component. But in a tree-local multicomponent TAG, the former would not be possible because the two adjunction sites are in different components.

We may generalize this concept to sLMGs in general.

Definition 27. A *multicomponent predicate* is one whose arguments are partitioned into one or more *components* (shown separated by a colon):

$$X(\alpha_{11},\ldots,\alpha_{1m_1} : \ldots : \alpha_{n1},\ldots,\alpha_{nm_n})$$

The *dissolution* of a multicomponent predicate with the above form is the set

$$\{X_1(\alpha_{11},\ldots,\alpha_{1m_1}),\ldots,X_n(\alpha_{n1},\ldots,\alpha_{nm_n})\}$$

Definition 28. A *set-local* (or *component-local*) *multicomponent production* is an sLMG production whose predicates are multicomponent predicates, and if two variables appear in the same component (or predicate, respectively) of the right-hand side, then they appear in the same component of the left-hand side.

The *dissolution* of a multicomponent production is the set of all possible well-formed sLMG productions formed out of the dissolutions of its predicates (keeping left-hand-side predicates on the left-hand side and right-hand-side predicates on the right-hand side).

For example, the elementary tree sets in Figure 2.4, with the adjunctions described above, could be converted into the following productions:

$$YZ(ax_1bx_2c, dx_3ex_4f : g,h) :\!- XX(x_1,x_4 : x_2,x_3) \qquad (\beta_1)$$
$$XX(ax_1b, cx_2d : ex_3f, gx_4h) :\!- YZ(x_1,x_2 : x_3,x_4) \qquad (\beta_2)$$

Both are easily verified to be set-local, but only β_1 is component-local: in β_2, x_1 and x_3 occur in the same right-hand-side predicate, they occur in different left-hand-side components.

Definition 29. A *set-local* (or *component-local*) *multicomponent sLMG* is an sLMG with set-local (or component-local, respectively) multicomponent productions and a single-component start predicate.

If a formalism $\mathscr{F}$ can be embedded in sLMG, then *set-local* (or *component-local*) *multicomponent* $\mathscr{F}$ consists of set-local (or component-local, respectively) multicomponent sLMGs such that the dissolution of each production is a well-formed production of $\mathscr{F}$.

Component-local multicomponent TAG is the same as tree-local multicomponent TAG. Set-local multicomponent CFG is also known as *local scattered-context grammar* [102].

Proposition 1. *If a formalism $\mathscr{F}$ can be embedded in sLMG, then component-local multicomponent $\mathscr{F}$ is weakly equivalent to $\mathscr{F}$.*

Proof. Observe that in the dissolution of a component-local multicomponent production, all the components of each right-hand-side predicate end up in the same production. Therefore, given a component-local multicomponent sLMG G, we can dissolve G and augment the nonterminal alphabet to guarantee that it has the same behavior as the original grammar. Let P' be the set of all productions that can be obtained as follows: for any $\pi \in P$, relabel the left-hand predicate to be π itself and relabel each right-hand predicate to a production with a matching left-hand side. Then let P'' be the union of the dissolutions of all the productions in P'. This set of productions forms a grammar of $\mathscr{F}$ weakly equivalent to G.

The component-locality constraint can be relaxed to *delayed component-locality* [38] without losing weak equivalence. Delayed tree-local multicomponent TAG has been employed for a variety of linguistic analyses [38, 58, 126].

2.5.3 Parsing

The basic parsing algorithm for sLMGs is straightforward: since they are just deductive systems, a chart-based deductive parser [125] can operate on an sLMG fairly transparently. Such a parser essentially searches the space of all possible instantiations of the productions of the input grammar for an instantiation with left-hand side $S(w)$, where w is the input string. Because the variables of an sLMG, as in an RCG, can only be instantiated to substrings of w, the number of instantiations of each sLMG production, and therefore the running time of the parser, is polynomial in $|w|$.

To obtain a parse forest, that is, a representation of all possible parses of a given string, we can use a construction due to Bertsch and Nederhof [12], a generalization of similar constructions for other formalisms [10, 128]. Given an sLMG G and an input string w, it computes another sLMG which compactly represents all possible derivations of w by G. (If G is linear, then the construction works for a general finite-state automaton, as shown by Bertsch and Nederhof.)

Define a new nonterminal alphabet

$$N' = \bigcup_{X \in N, m \leq f} \{X\} \times Q^{2m}$$

where $Q = \{0, \ldots, |w|\}$ and f is the maximum arity of any predicate. Intuitively, each nonterminal $X \in N$ and arity m is accompanied by $2m$ string positions, indicating the left and right input positions of the arguments of X.

For any pair of mappings $q, r : V \to Q$ (where $q(x) \leq r(x)$) and string $\alpha \in (V \cup T)^*$, define the condition $match_{q,r}(\alpha, w, i, j)$ which is true if one of the following is true:

- $\alpha = w_{i+1}\beta$ and $match_{q,r}(\beta, w, i+1, j)$
- $\alpha = x\beta$, $x \in V$, $q(x) = i$, and $match_{q,r}(\beta, w, r(x), j)$
- $\alpha = \varepsilon$ and $i = j$

Roughly speaking, $match_{q,r}(\alpha, w, i, j)$ holds just in case making each variable x to span $w_{q(x)+1} \cdots w_{r(x)}$ makes α to span $w_{i+1} \cdots w_j$.

Define a new set of productions P' as follows. For each production $\pi \in P$ of the form

$$X(\alpha_1, \ldots, \alpha_m) :- Y_1(\beta_{11}, \ldots, \beta_{1m_1}), \ldots, Y_n(\beta_{n1}, \ldots, \beta_{nm_n})$$

and for every pair of mappings $q, r : V \to Q$, if for each α_i there exist $q_i, r_i \in Q$ such that $match_{q,r}(\alpha_i, w, q_i, r_i)$, then add to P' the production

$$X'(\alpha_1, \ldots, \alpha_m) :- Y_1'(\beta_{11}, \ldots, \beta_{1m_1}), \ldots, Y_n'(\beta_{n1}, \ldots, \beta_{nm_n})$$

where $X' = \langle X, q_1, r_1, \ldots, q_m, r_m \rangle$ and $Y_i' = \langle Y_i, q(\beta_{i1}), r(\beta_{i1}), \ldots, q(\beta_{im_i}), r(\beta_{im_i}) \rangle$.

These productions form a new grammar G_w whose start symbol is $\langle S, 0, |w| \rangle$. Observe that even if G was nonlinear, G_w is essentially a CFG; the nonterminals do all the work and the arguments do not further constrain the derivations. The size of G_w is $\mathcal{O}(|G|n^{(r+1)f})$ productions, where $|G|$ is the number of productions, n is the length of w, f is the maximum arity, and r is the maximum number of nonterminals on the right-hand side of a production. Therefore the running time of this construction is also $\mathcal{O}(|G|n^{(r+1)f})$.

This algorithm is by no means optimal, however. In order to improve parsing time, one would have to write a specialized parser for the grammar or, equivalently, construct a *cover grammar* for it. For example, in order to achieve $\mathcal{O}(n^3)$ time complexity for parsing CFGs, we must either convert the CFG into Chomsky normal form (that is, strict binary-branching) or use a specialized algorithm like Earley's algorithm, which effectively binarizes the grammar on the fly. In either case, there must be a way of reconstructing the derivations of the original grammar. We discuss this topic further below.

2.5.4 Cover Grammars

When considering the use of new formalisms, we are especially interested in maximizing a formalism's power in one respect while minimizing it in another. Most often, one wants to minimize its WGC: Joshi [68] speaks of "squeezing" SGC out of a formalism without increasing its WGC. From a theoretical standpoint, such formalisms are interesting because they point to finer-grained ways of measuring formal power than the traditional measure of WGC. More practically, one may want to minimize parsing complexity: it would be ideal to gain extra SGC without increasing the asymptotic complexity of the parsing algorithm.

These two constraints, WGC and parsing complexity, often coincide: proofs of weak equivalence to, say, CFG, are often accompanied by $\mathcal{O}(n^3)$ parsing algorithms.

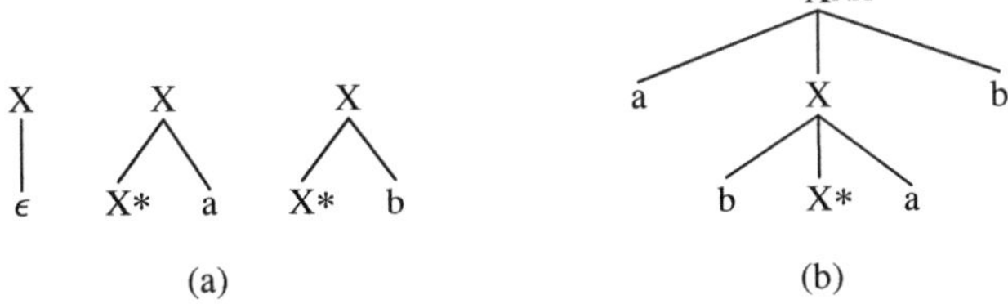

Fig. 2.5 Example of weakly context-free TAG.

Indeed, we argue that to have one without the other is not likely to be very interesting.[3] For example, the TAG of Figure 2.5a generates a CFL ($\{a,b\}^*$), and adding the tree in Figure 2.5b does not increase the grammar's WGC, yet intuitively it increases the grammar's complexity. Conversely, the language

$$\{a^i b^j c^k \mid ijk + 1 \text{ is a prime number}\}$$

has a (very naïve) $\mathscr{O}(n^3)$ recognition algorithm, but whatever formal system generates it is not likely to resemble a CFG.

It would be easier to characterize what it means to squeeze SGC out of a formalism if we used a tighter constraint, one which entailed both preservation of WGC and preservation of parsing complexity. Such a constraint is suggested by examining the parsing algorithms for common weakly context-free formalisms. For example, the parsers for RF-TAG and TIG are based on CKY; their items are of the form $[X, i, j]$ and are combined in various ways, but always according to the deductive rule schema

$$\frac{[Y, i, j] \quad [Z, j, k]}{[X, i, k]}$$

where the material below the line is deduced from the material above the line [125]. But this is just like the CKY parser for CFG in Chomsky normal form. In effect the parser dynamically converts the RF-TAG or TIG into an equivalent Chomsky-normal-form CFG—each parser rule of the above form corresponds to the rule schema $X \rightarrow YZ$.

More importantly, given a grammar G and a string w, a parser can reconstruct a packed forest of all possible derivations of w in G by storing some information inside its chart items. Every time it generates a new item, it takes the derivation information in the antecedent items to compute some new information for the new item. If we think of the parser as dynamically converting G into a CFG G', then we may think of these computations as attached not to the deductive rules of the parser, but to the productions of G'. Indeed, we may think of them as a kind of interpretation function for G' into the domain of G-derivations. We call G' a *cover grammar* for G, following Nijholt [94]. This covering relationship is a relationship between

---

grammars and not parsing strategies: while this covering relationship prescribes a particular approach to parsing G, it is independent of any particular parsing strategy for G'.

So far we have not articulated how the derivations of G are reconstructed. A parser like CKY builds a packed parse forest by storing in each parser item $[X, i, j]$ a set of productions $X \rightarrow YZ$ together with *back-pointers* to forests of derivations of Y and Z. Crucially, because these subforests would be exponential in size if unpacked, the parser never accesses their internals; it only deals with back-pointers to them, without dereferencing them. Parsers which use cover grammars also typically use back-pointers in this way, which we formalize as follows:[4]

Definition 30. A *cover sLMG G'* is an sLMG together with an interpretation function whose local interpretation functions f_π operate on tuples of derivations and are each definable as:

$$f_\pi(\langle t_{11}, \ldots, t_{1m_1} \rangle, \ldots, \langle t_{n1}, \ldots, t_{nm_n} \rangle) = \langle u_1, \ldots, u_m \rangle$$

where the t_{ij} are variables and each u_i is drawn from the set τ, which is a set of derivation fragments recursively defined as follows:

1. t_{ij} is in τ (copying a back-pointer)
2. $\pi'(\tau_1, \ldots, \tau_n)$ is in τ, where $\tau_i \in \tau$ and π' is an sLMG production (creating a derivation fragment with back-pointers)

Definition 31. We say that a cover sLMG G' *covers* another sLMG G if there is a one-to-one correspondence between derivations of G' and derivations of G such that G' derives w with interpretation δ if and only if δ is the corresponding G-derivation, and δ is a derivation of w. We say that the cover *respects* an interpretation domain D if corresponding derivations also have the same interpretation in D.

Definition 32. We say that a formalism $\mathscr{F}'$ *covers* another formalism $\mathscr{F}$ (respecting D) if for any grammar G provided by $\mathscr{F}$, there is a grammar provided by $\mathscr{F}'$ which can cover G (respecting D).

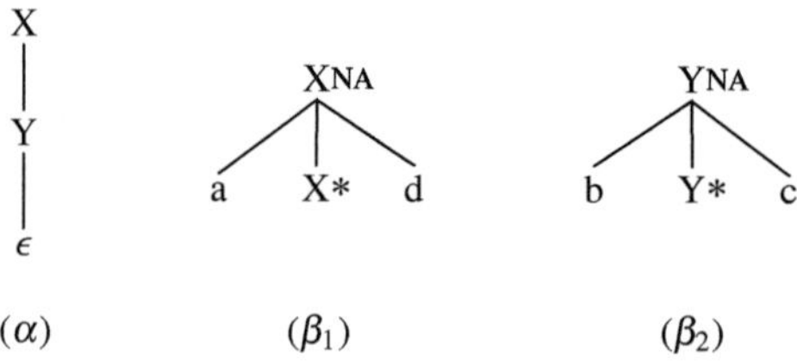

Fig. 2.6 Example TAG to be covered. Here adjunction at foot nodes is allowed.

---

[4] This notion is similar to generalized syntax-directed translation [5].

	Production	Name
	$S(x_1 y_1 y_2 x_2) :- X(x_1, x_2), Y(y_1, y_2)$	(α)
	$X(ax_1, x_2 d) :- X(x_1, x_2)$	(β_1)
	$X(\varepsilon, \varepsilon)$	(ε_X)
	$Y(by_1, y_2 c) :- Y(y_1, y_2)$	(β_2)
	$Y(\varepsilon, \varepsilon)$	(ε_Y)

Fig. 2.7 sLMG representation of the TAG of Figure 2.6.

Production	Local interpretation function	Comment
$S \to \alpha^{0\bullet}$	$\alpha(t_1, t_2) \leftarrowtail \langle t_1, t_2 \rangle$	
$\alpha^{0\bullet} \to \alpha^0_\bullet$	$\langle \varepsilon_X(), t_2 \rangle \leftarrowtail \langle -, t_2 \rangle$	no adjunction
$\alpha^0_\bullet \to \alpha^{1\bullet}$	$\langle -, t_2 \rangle \leftarrowtail \langle -, t_2 \rangle$	
$\alpha^{1\bullet} \to \alpha^1_\bullet$	$\langle -, \varepsilon_Y() \rangle \leftarrowtail \langle -, - \rangle$	no adjunction
$\alpha^1_\bullet \to \varepsilon$	$\langle -, - \rangle$	
$\alpha^{0\bullet} \to \beta_1^0[\alpha^0]$	$\langle \beta_1(t_1), t_2 \rangle \leftarrowtail \langle t_1, t_2 \rangle$	adjoin β_1
$\beta_1^0[\alpha^0] \to a\,\beta_1^2[\alpha^0]\,d$	$\langle t_1, t_2 \rangle \leftarrowtail \langle t_1, t_2 \rangle$	
$\beta_1^2[\alpha^0] \to \beta_1^0[\alpha^0]$	$\langle \beta_1(t_1), t_2 \rangle \leftarrowtail \langle t_1, t_2 \rangle$	adjoin β_1 by "tail recursion"
$\beta_1^2[\alpha^0] \to \alpha^0_\bullet$	$\langle \varepsilon_X(), t_2 \rangle \leftarrowtail \langle -, t_2 \rangle$	no adjunction, return to α
$\alpha^{1\bullet} \to \beta_2^0[\alpha^1]$	$\langle -, \beta_2(t_2) \rangle \leftarrowtail \langle -, t_2 \rangle$	adjoin β_2
$\beta_2^0[\alpha^1] \to b\,\beta_2^2[\alpha^1]\,c$	$\langle -, t_2 \rangle \leftarrowtail \langle -, t_2 \rangle$	
$\beta_2^2[\alpha^1] \to \beta_2^1[\alpha^1]$	$\langle -, \beta_2(t_2) \rangle \leftarrowtail \langle -, t_2 \rangle$	adjoin β_2 by "tail recursion"
$\beta_2^2[\alpha^1] \to \alpha^1_\bullet$	$\langle -, \varepsilon_Y() \rangle \leftarrowtail \langle -, - \rangle$	no adjunction, return to α

Fig. 2.8 CFG cover of the TAG of Figure 2.6. Here we leave the local interpretation functions anonymous; $y \leftarrowtail x$ denotes the function which maps x to y.

As an example, a CFG which covers the RF-TAG of Figure 2.6 is shown in Figure 2.8. The nonterminals of this grammar consist of an elementary tree name, a superscripted tree address, a dot indicating the "top half" or "bottom half" of the node (to prevent multiple adjunctions at a node), and a stack in square brackets. When a tree β is adjoined into another tree γ, γ is pushed onto the stack so that it can be recalled when β is finished; however, if β is adjoined at the foot node of γ, then γ does not need to be recalled, so it is not pushed onto the stack, as in the programming-language technique of tail recursion.

The notion of a cover grammar provides a new view of the question posed by Joshi [68], "How much strong generative power can be squeezed out of a formal system without increasing its weak generative power?" In our present framework, we must understand SGC to be relative to some interpretation domain. Moreover, in light of the foregoing arguments, we should modify the constraint on WGC to be a constraint on coverability. This provides a more rigid framework in which Joshi's question can be explored.

We will show in later chapters that in some interpretation domains, formalisms that are coverable by CFG can indeed have greater SGC than CFG. For example, in Section 5.3.1 we show that RF-TAG can generate a linked string set that CFG cannot:

$$L = \left\{ c\,a\,a\,\cdots\,a\,c\,b\,\cdots\,b\,b \right\}$$

In a sense, this greater SGC is squeezed out of CFG for free. But this kind of squeezing has its limits.

Proposition 2. *If G' covers G (using interpretation function c), then for any interpretation function f of G, there exists an equivalent interpretation function f' of G', that is, $f' = f \circ c$.*

Proof. The equivalent interpretation for G' is easy to construct: the basic idea is that wherever a local interpretation function in the cover generates a G-production π, we substitute π's local interpretation function in place of π.

This means that we could in fact construct an interpretation for the cover CFG of Figure 2.8 that generates L. It is only under the restriction that links be defined within local domains that L is impossible for CFG.

In Proposition 2, the interpretation of G could even be another cover, which implies that coverability is transitive. This means that while a covered formalism might have greater SGC than the cover formalism in some domains, it can never have greater SGC in the domain of covered derivations.

Corollary 1. *A formalism $\mathscr{F}'$ can cover another formalism $\mathscr{F}$ if and only if $\mathscr{F}'$ can cover every sLMG that $\mathscr{F}$ can.*

In other words, one cannot squeeze a formalism a second time to get still more power out. Therefore the class of sLMGs coverable by CFG represents the maximum amount of SGC that can be squeezed out of CFG as we have defined it.

In an earlier paper [32] we tried to characterize this class of grammars more directly by choosing an interpretation domain and exhibiting a formalism that maximized SGC with respect to this domain. But since there are many different ways to do this, it is more fruitful to consider Joshi's question with reference to a particular application.

Chapter 3
Statistical Parsing

We now turn to *statistical parsing*, in which some probability model defined over parse structures (standardly, phrase-structure trees) is used to determine the best structure or best k structures of a given string. We introduce *weighted* interpretation domains and show how parsing models of a very general nature can be expressed as weighted grammars. But we find that this domain classifies formalisms rather coarsely: any two formalisms with the same coverability (respecting parse structures) also define the same parsing models.

Though this result makes the hope rather dim of squeezing statistical-modeling power out of PCFG for free, it invites a reinterpretation of lexicalized PCFG models [25, 44] as cover grammars of grammars with richer structural descriptions than phrase-structure trees. As a demonstration of this new view, we define a probabilistic TAG model and discuss techniques for obtaining adequate models from corpora which, from this point of view, are labeled only with partial structural descriptions. This model performs at the same level as lexicalized PCFG parsers and captures the same kinds of dependencies they do in a conceptually simpler way.

3.1 Measuring Statistical Modeling Power

Typically, grammar-based statistical parsing models are defined as weighted grammars, which we may define as follows.

Definition 33. The interpretation domain of *weights* is the set of nonnegative reals; its local interpretation functions are of the form

$$f_\pi(w_1, \ldots, w_n) = w \times \prod_{i=1}^{n} w_i \tag{3.1}$$

A *weighted sLMG* is an sLMG together with an interpretation function for the domain of weights.

In order to compare parsing models across formalisms (in terms of SGC or in terms of parsing accuracy), we need to map derivations to some common parse structure, usually phrase-structure trees or dependency trees. Here, we only consider phrase-structure trees. Thus, we measure SGC in the domain of weighted trees (the domain whose interpretations are (tree, weight) pairs and whose interpretation functions are each the product of an interpretation function for the domain of trees and an interpretation function for the domain of weights).

The statistical parsing problem is then to compute the highest-weighted tree for a given input sentence. If it is possible for more than one derivation to yield the same tree, a question arises: should we look for the highest-weight derivation or the highest-weight tree (where the weight of a tree is the sum of the weights of the derivations that yield it)? Some approaches to statistical parsing, like data-oriented parsing [17], do the latter. But most parsers simply look for the highest-weight derivation and output its interpretation as a tree.

The weights need not be numeric. If we want to estimate weights from data, then we must start with a grammar whose productions π each have for a weight a *feature vector* $\mathbf{h}(\pi)$. The feature vector of a derivation is the sum of the feature vectors of its productions. To train the model, we estimate a vector of *feature weights* $\mathbf{w}$; this allows us to assign the numeric weight $\mathbf{w} \cdot \mathbf{h}(\pi)$ to each production π.

For example, a linear sLMG G can be made into a probabilistic version by assigning a different unit feature vector (that is, a vector with value 1 for one feature and 0 for the rest) to each production. The feature-weight vector to be estimated is a vector of log-probabilities. For each nonterminal X and arity n, let $P_{X,n}$ be the set of productions whose left-hand side has nonterminal X and arity n. Then we require

$$\sum_{\pi \in P_{X,n}} \exp(\mathbf{w} \cdot \mathbf{h}(\pi)) = 1 \tag{3.2}$$

Subject to this constraint, the maximum-likelihood estimate for $\mathbf{w}$ is just

$$\exp(\mathbf{w} \cdot \mathbf{h}(\pi)) = \frac{c(\pi)}{\sum_{\pi \in P_{X,n}} c(\pi')} \tag{3.3}$$

where $c(\pi)$ is the number of occurrences of π in the training data. The reason G must be linear is that otherwise the derivation distribution would not sum to one in general. This schema applied to CFG gives probabilistic CFG [18]. Probabilistic TAG [105, 112] can be constructed in a similar way, although the details depend on how we embed TAG into sLMG.

As another example, we can let the production weights be arbitrary feature vectors, and instead of the constraint (3.2), we simply renormalize the derivation distribution; that is, if $\mathcal{D}$ is the set of derivations of G and $d \in \mathcal{D}$, then

$$P(d) = \frac{\exp(\mathbf{w} \cdot \sum_{\pi \in d} \mathbf{h}(\pi))}{\sum_{d' \in \mathcal{D}} \exp(\mathbf{w} \cdot \sum_{\pi \in d'} \mathbf{h}(\pi))} \tag{3.4}$$

This is applicable even to nonlinear sLMGs. It defines a log-linear (or maximum-entropy) model over derivations whose features are local to a production, but can be arbitrary functions on productions. Since the summation in the denominator can be difficult to compute [3], it is more common to restrict it to derivations of a given string w, so that the model defines a conditional distribution $P(d \mid w)$ [66]. Estimating the feature weights $\mathbf{w}$ can be done by various numerical methods, which generally involve calculating the expected number of times that a feature occurs for a given training example. This can be done using a generalization of the Inside-Outside algorithm [9, 79, 43, 91, 34].

In order to measure the statistical-modeling power of a grammar formalism, then, we should look at its SGC with respect to the domain of weighted trees, whether those weights are numbers or feature vectors.

Can we squeeze statistical-modeling power out of a formalism for free—that is, without affecting coverability? The answer is the following negative result:

Proposition 3. *If a grammar G' covers G respecting some interpretation domain D, then the weighted versions of G' and G are equivalent with respect to the weighted version of D.*

Proof. Let G_w be the weighted version of G. Given G', it is easy to construct a weighted version of G' that is equivalent to G_w: replace each local interpretation function f' in G' by (3.1), with p equal to the product of the weights in G_w of the local interpretation functions invoked by f'. The resulting weighted grammar G'_w is equivalent to G_w, for even though G'_w and G_w multiply the production weights together in different orders, the result is the same because multiplication of weights (or addition of feature vectors) is commutative.

This means that squeezing SGC while preserving coverability will not increase the number of models that can be described. On the other hand, it is trivially true that a formalism with greater SGC with respect to parse structures can describe more models. For example, TIG (see Section 2.5.1), because it can be covered by CFG respecting strings, cannot be used to describe any more models of string probability than CFG can. But because it has greater tree generative capacity than CFG, it can describe more models of tree probability. Therefore, since many recent parsing models [25, 44, 90] are based on PCFG, we should not expect to improve much on them by moving to TIG, and not at all by moving to TSG or RF-TAG.

3.2 Lexicalized Probabilistic CFG

However, this result reveals a deeper insight into these models. The CFGs they are based on are called *lexicalized*, which means that they have been modified so that every nonterminal symbol contains a lexical item (see Figure 3.1 for an example derivation). Before training, a set of rules [86] is used to identify a *head child* for every node in the training data; then the training data are transformed in the following way, bottom-up:

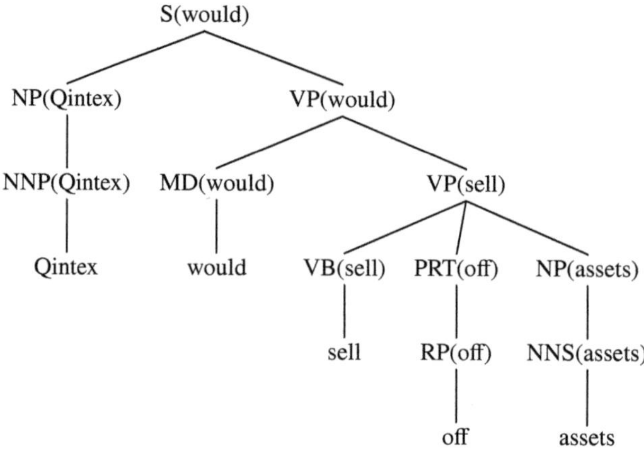

Fig. 3.1 Lexicalized PCFG tree.

1. If a node labeled X has a head child which is a terminal node labeled w, then relabel the node $X(w)$.
2. If a node labeled X has a head child labeled $Y(w)$, relabel the node $X(w)$.

The conventional wisdom regarding lexicalized PCFG is that it rearranges the lexical information in trees so that it can be used more effectively; specifically, it places pairs of words into the same local domain so that *bilexical statistics* can be collected. But experiments have shown [54, 72, 13] that bilexical statistics actually help very little in parsing. We argue below that the lexicalization process does more than rearrange lexical information to create bilexical dependencies.

Recall that a cover grammar has pieces of the covered derivations attached to its productions, and it uses information transmitted through its nonterminal symbols to ensure that the pieces are attached correctly. For example, one way to construct a cover grammar for a TSG (see Section 2.5.1) could be:

1. For every non-substitution node labeled X with address η in an elementary tree α, add as a decoration the singleton set

$$\{\langle \alpha, \eta \rangle\}$$

2. For every substitution node labeled X, add the decoration

$$\{\langle \alpha, \varepsilon \rangle \mid \alpha \text{ is an elementary tree with root label } X\}$$

3. For every node labeled X with decoration $\{\langle \alpha, \eta \rangle\}$ immediately dominating nodes labeled $X_1, \ldots, X_n$ with decoration $\Delta_1, \ldots, \Delta_n$, construct the CFG rules

$$X(\alpha, \eta) \rightarrow X_1(\alpha_1, \eta_1), \ldots, X_n(\alpha_n, \eta_n) \qquad \text{for all } \langle \alpha_i, \eta_i \rangle \in \Delta_i$$

If we further assume that each elementary tree α has a single lexical anchor w_α, then observe that (α, η) subsumes (w_α), so that the CFG with rules

$$X(w_\alpha) \to X_1(w_{\alpha_1}), \ldots, X_n(w_{\alpha_n}) \qquad \text{for all } \langle \alpha_i, \eta_i \rangle \in \Delta_i$$

generates an approximate superset of the original TSG. But this grammar is none other than a lexicalized PCFG. We may therefore think of a lexicalized PCFG as an approximate cover of a TSG. If each elementary tree has a unique lexical anchor and each node in a tree has a unique label, then the cover is exact.

Collins' models decompose the generation of productions more finely; we omit the details here, only noting that the use of a Markov process to generate adjuncts makes an infinite number of productions possible. To avoid infinite grammars, we add sister-adjunction (see Section 2.5.1), which can create new children under a node in a manner similar to Collins' Markov model.

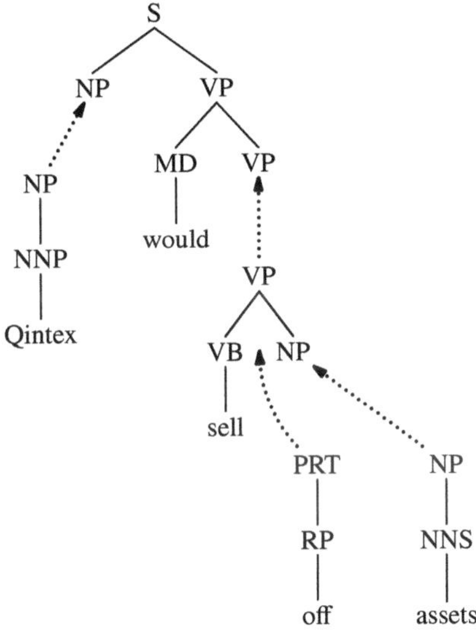

Fig. 3.2 Derivation of TSG with sister-adjunction corresponding to tree of Figure 3.1.

Therefore Collins' Model 2, minus its distance measure, can alternatively be thought of as defined on a TSG with sister-adjunction (see Figure 3.2). Since a TSG's derivations are distinct from its derived trees and contain more information than them, we should likewise think of Collins' Model 2 as being defined over richer structural descriptions than are found in the Penn Treebank, and we should think of the lexicalization process as reconstructing information rather than rearranging information, and this information is structural rather than lexical.

In the following section we build a parsing model from the ground up according to this new perspective. This perspective also suggests an alternative to reconstruc-

tion heuristics: to treat training as a partially-unsupervised learning problem and use
EM to train the model from partial structural descriptions.

3.3 A Probabilistic TIG Model

If lexicalized CFG is a cover grammar for something like a TAG, then perhaps a
statistical parsing model should be defined directly as one. In this section we define a
model based on probabilistic TIG with sister-adjunction. Such a formulation would
have several advantages:

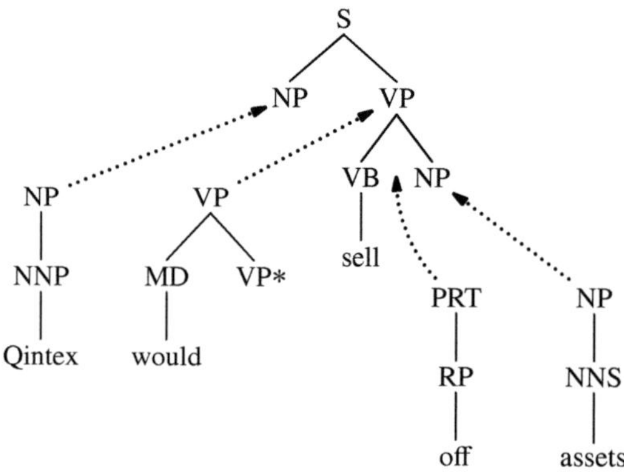

Fig. 3.3 Derivation of TIG with sister-adjunction corresponding to tree of Figure 3.1.

First, we noted that lexicalized CFG is only an approximate cover because it uses
lexical anchors as a proxy for their elementary structures. Various modifications to
lexicalized CFG have been found to improve parsing accuracy, for example:

- an S which does not dominate an argument NP to the left of its head is relabeled
  SG, so that the attachment of the clause can be conditioned on whether it has a
  subject [45]
- an NP which does not dominate another NP which does not dominate a POS is
  relabeled NPB, and if its parent is not an NP, an NP node is inserted [45] because
  PCFG mismodels Treebank-style two-level NPs [65]
- every node's label is augmented with the label of its parent [26, 65]

Such changes are not always obvious *a priori* and often must be devised anew for
each language or each corpus. But the above modifications are not necessary in a
TAG-like model, because it has statistics of pairs of elementary trees and not just
pairs of words. Thus many dependencies that have to be stipulated in a PCFG by
tree transformations are captured for free in a probabilistic TAG model.

Second, TAG provides greater flexibility in defining heuristics. For example, we
might want elementary trees that contain both a preposition and the head word of

the prepositional object, in the hope that the latter will help make PP attachment decisions [46]. Or, in a sentence with an auxiliary verb, like the above example "Qintex would sell off assets," we might want the subject to attach to the tree for 'sell' instead of the tree for 'would.' It would be tricky to make such changes to a head word percolation scheme, but easy with a TAG or TIG (see Figure 3.3 for a TIG derivation of the latter example).

Third, by decoupling the reconstruction heuristics from the training process proper, this new view suggests alternative training methods. Below we describe experiments using Expectation-Maximization to train directly on the observed, not the reconstructed data (a method explored previously by Hwa [63] for TIG).

Basic definition

The parameters of a probabilistic TAG model [105, 112] are:

$$\sum_{\alpha} P_i(\alpha) = 1$$

$$\sum_{\alpha} P_s(\alpha \mid \eta) = 1$$

$$\sum_{\beta} P_a(\beta \mid \eta) + P_a(\text{NONE} \mid \eta) = 1$$

where α ranges over initial trees, β over auxiliary trees, and η over nodes. $P_i(\alpha)$ is the probability of beginning a derivation with α; $P_s(\alpha \mid \eta)$ is the probability of substituting α at η; $P_a(\beta \mid \eta)$ is the probability of adjoining β at η; finally, $P_a(\text{NONE} \mid \eta)$ is the probability of nothing adjoining at η. The probability of a derivation can then be expressed as a product of the probabilities of the individual operations of the derivation.

We restrict the TAG to be a TIG for efficiency reasons. The above model works for TIG just as it does for TAG. However, the original definition of probabilistic TIG [117] is flawed because it allows one left auxiliary tree and one right auxiliary tree (but not more than one of each) to adjoin at the same node in either order, but the probability model does not distinguish the two orders, so that the total probability of all valid derivations is greater than one. Hwa [64, p. 30] describes how to fix the problem, but our fix is simply to prohibit this type of simultaneous adjunction.

Our variant (henceforth TIG-SA) adds another set of parameters for sister-adjunction:

$$\sum_{\alpha} P_{sa}(\alpha \mid \eta, i, \alpha') + P_{sa}(\text{STOP} \mid \eta, i, \alpha') = 1$$

where α and α' range over initial trees, and (η, i) over possible sister-adjunction sites. Let n be the number of children of η; $P_{sa}(\alpha \mid \eta, i, \alpha')$, $0 \leq i \leq n$, is the probability of sister-adjoining α under η at position i, when α' is the previous[1] tree to have sister-adjoined at that position (or START if none). Thus modifier trees are generated as a first-order Markov process, as in the model used in BBN's SIFT system [90] and the base-NP model of Collins [45].

Independence assumptions and smoothing

Since the number of parameters in this model is too high to get reasonable estimates from corpus data, we generate an elementary tree in two steps and smooth each step: first the *tree template* (that is, the elementary tree stripped of its anchor), then the anchor. Thus:

$$P_i(\alpha) = P_{i_1}(\tau_\alpha)P_2(w_\alpha \mid \tau_\alpha)$$
$$P_s(\alpha \mid \eta) = P_{s_1}(\tau_\alpha \mid \eta) \cdot P_2(w_\alpha \mid \tau_\alpha, t_\eta, w_\eta)$$
$$P_a(\beta \mid \eta) = P_{a_1}(\tau_\beta \mid \eta) \cdot P_2(w_\beta \mid \tau_\beta, t_\eta, w_\eta)$$
$$P_{sa}(\alpha \mid \eta, i, \alpha') = P_{sa_1}(\tau_\alpha \mid \eta, i, X_{\alpha'}) \cdot P_2(w_\alpha \mid \tau_\alpha, t_\eta, w_\eta, X_{\alpha'})$$

where τ_α is the tree template of α and w_α is the lexical anchor of α, and similarly for β; w_η is the lexical anchor of the elementary tree containing η, and t_η is the part-of-speech tag of that anchor. We have reduced α' to its root label $X_{\alpha'}$. Note that the same probability P_2 is used for all three composition operations: for adjunction and substitution, $X_{\alpha'}$ is assigned the value START.

These probabilities each have three backoff levels:

	$P_{s_1,a_1}(\gamma \mid \cdots)$	$P_{sa_1}(\alpha \mid \cdots)$	$P_2(w \mid \cdots)$
1	$\tau_\eta, w_\eta, \eta_\eta$	$\tau_\eta, w_\eta, \eta_\eta, i, X_{\alpha'}$	$\tau_\gamma, t_\eta, w_\eta, X_{\alpha'}$
2	τ_η, η_η	$\tau_\eta, \eta_\eta, i, X_{\alpha'}$	$\tau_\gamma, t_\eta, X_{\alpha'}$
3	$\bar{\tau}_\eta, \eta_\eta$	$\bar{\tau}_\eta, \eta_\eta, i$	τ_γ
4	X_η	$\emptyset$	t_γ

where τ_η is the tree template of the elementary tree containing η, $\bar{\tau}_\eta$ is τ_η stripped of its anchor's POS tag, X_η is the label of η, and η_η is the address of η in its elementary tree; τ_γ is the tree template of γ, and t_γ is the POS tag of its anchor. The backed-off models are combined by linear interpolation:

$$e = \lambda_1 e_1 + (1 - \lambda_1)(\lambda_2 e_2 + (1 - \lambda_2)(\lambda_3 e_3 + (1 - \lambda_3)e_4)) \tag{3.5}$$

where e_i is the estimate at level i, and the λ_i are computed by a combination of formulas used by Collins [45] and Bikel et al. [16]:

---

[1] Here "previous" means "next closest to the lexical anchor," which presupposes a single lexical anchor; we could alternatively define it to mean "next to the left," which would be more general but less efficient.

$$\lambda_i = \left(1 - \frac{d_{i-1}}{d_i}\right)\left(\frac{1}{1 + 5u_i/d_i}\right) \qquad (3.6)$$

where d_i is the number of occurrences in training of the context at level i ($d_0 = 0$), and u_i is the number of unique outcomes for that context seen in training.

To handle unknown words, we treat all words seen five or fewer times in training as a single symbol UNKNOWN, following Collins [44].

Parsing

We used a CKY-style TIG parser similar to the one described by Schabes and Waters [117], with a modification to ensure completeness (because foot nodes are effectively empty, which standard CKY does not handle). We also extended the parser to allow sister-adjunction.

The parser uses a beam search, assigning a score to each item $[\eta, i, j]$ and pruning away any item with score less than 10^{-5} times that of the best item for that span, following Collins [44]. The score of an item is its inside probability multiplied by the prior probability $P(\eta)$, following Goodman [56]. $P(\eta)$, in turn, is decomposed as $P(\bar{\tau}_\eta \mid t_\eta, w_\eta) \cdot P(t_\eta, w_\eta)$, so that the first term can be smoothed by linear interpolation (as above) with the backed-off estimate $P(\bar{\tau}_\eta \mid t_\eta)$, again following Collins [45].

As mentioned above, words occurring five or fewer times in training were replaced with the symbol UNKNOWN. When any such word w occurs in the test data, it is also replaced with UNKNOWN. Following Collins [44], the parser only allows w to anchor templates that have POS tags observed in training with w itself, or templates that have the POS tag assigned to w by MXPOST [104]; all other templates are thrown out for w.

When parsing English, we use Collins' comma rule: when the parser combines two constituents, if the right-hand constituent has a comma to its left, it must also have a comma (or the end of the sentence) to its right, or else the two constituents cannot be combined. In our parser, because a binary-branching cover grammar is used, this means that if a right modifier (substitution or sister-adjunction) has a comma to its left, it must have a comma (or the end of the sentence) to its right; if a left modifier has a comma to its left, then the *parent node* must have a comma to its right.

3.4 Training from Partial Structural Descriptions

We want to obtain a maximum-likelihood estimate of these parameters, but cannot estimate them directly from the Treebank, because the sample space of probabilistic TIG-SA is the space of TIG-SA derivations, not the derived trees that are found in the Treebank. For there are many TIG-SA derivations which can yield the same derived tree, even with respect to a single grammar. We need, then, to reconstruct TIG-SA derivations somehow from the training data.

One approach, taken by Magerman [86] and others for lexicalized PCFGs and Neumann [93] and others [132, 27] for TAGs, is to use heuristics to reconstruct the derivations, and directly estimate the probabilistic TIG-SA parameters from the reconstructed derivations. Another approach, taken by Hwa [63], is to choose some grammar general enough to parse the whole corpus and obtain a maximum-likelihood estimate by Expectation-Maximization. Below we discuss the first approach, and then a combination of the two approaches.

3.4.1 Rule-Based Reconstruction

Given a CFG and a Magerman-style head-percolation scheme, an equivalent TIG-SA can be constructed whose derivations mirror the dependency analysis implicit in the head-percolation scheme.

For each node η, the head-percolation and argument/adjunct rules classify exactly one child of η as a head and the rest as either arguments or adjuncts. We use Magerman and Collins' rules with few modifications (see Tables 3.4 and 3.5), but we treat coordination specially: if an X dominates a CC, and the rightmost CC has an X to its left and to its right, then that CC is marked as the head and the two nearest Xs on either side as arguments, and no further rules apply.

Using this classification into heads, arguments, and adjuncts, we can construct a TIG-SA derivation (including elementary trees) from a derived tree as follows:

1. If η is an adjunct, excise the subtree rooted at η to form a sister-adjoined initial tree.
2. If η is an argument, excise the subtree rooted at η to form an initial tree, leaving behind a substitution node.
3. But if η is an argument and η' is the nearest ancestor with the same label, and η is the rightmost descendant of η', and all the intervening nodes, including η', are heads, excise the part of the tree from η' down to η to form an auxiliary tree, leaving behind a head node.

Rules (1) and (2) produce the desired result; rule (3) changes things somewhat by making trees with recursive arguments into auxiliary trees. Its main effect is to extract VP auxiliary trees for modal and auxiliary verbs. In the present experiments, in fact, it is restricted to nodes labeled VP. The complicated restrictions on η' are simply to ensure that a well-formed TIG-SA derivation is produced.

When we run this algorithm on sections 02–21 of the Penn Treebank [87], the resulting grammar has 50,628 lexicalized trees (with words seen five or fewer times replaced with UNKNOWN). However, if we consider elementary tree templates, the grammar is quite manageable: 2104 tree templates, of which 1261 occur more than once (see Figure 3.5). A few of the most frequent tree templates are shown in Figure 3.4.

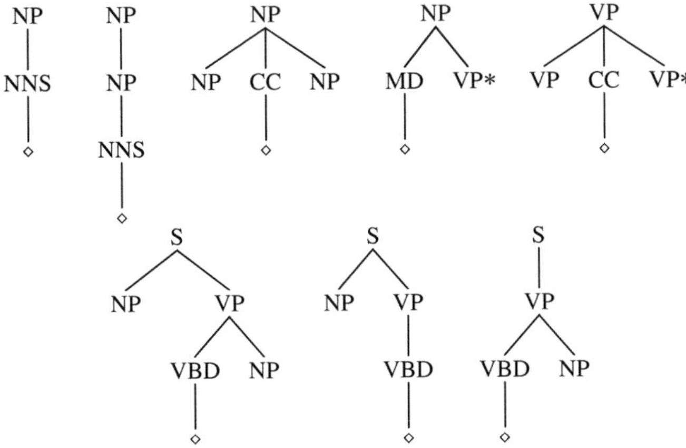

Fig. 3.4 Examples of frequently-occurring extracted tree templates. ⋄ marks where the lexical anchor is inserted.

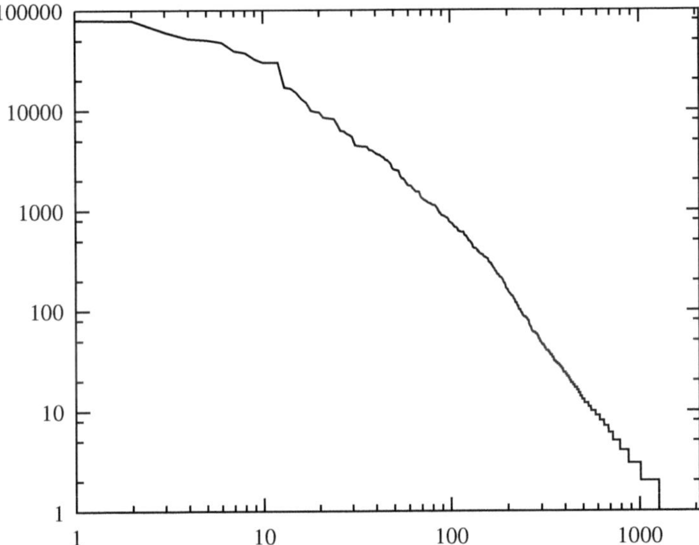

Fig. 3.5 Distribution of tree templates: frequency versus rank (log-log).

So the extracted grammar is fairly compact, but how complete is it? Ideally the size of the grammar would converge, but if we plot its growth during training (Figure 3.6), we see that even after training on 1 million words, new elementary tree templates continue to appear at the rate of about four every 1000 words, or one every ten sentences.

We do not consider this effect to be seriously detrimental to parsing. Since 90% of unseen sentences can be parsed perfectly with the extracted grammar, its cov-

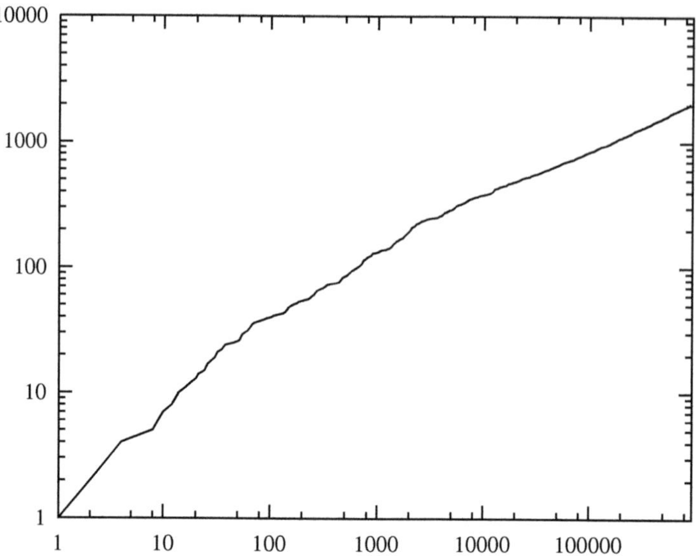

Fig. 3.6 Growth of grammar during training: types versus tokens (log-log).

erage is good enough potentially to parse new data with state-of-the-art accuracy. Note that even for the remaining 10% it is still quite possible for the grammar to derive a perfect parse, since there can be many TIG-SA derivations which yield the same derived tree. Nevertheless, we would like to know the source of this effect and minimize it. Three possible explanations are:

- New constructions continue to appear.
- Old constructions continue to be (erroneously) annotated in new ways.
- Old constructions continue to be combined in new ways, and the extraction heuristics fail to factor this variation out.

In a random sample of 100 once-seen elementary tree templates, we found (by casual inspection) that 34 resulted from annotation errors, 50 from deficiencies in the heuristics, and four apparently from errors in the text itself. Only twelve appeared to be genuine.[2]

Therefore the extracted grammar is more complete than Figure 3.6 suggests at first glance. Evidently, however, our extraction heuristics have room to improve. The majority of trees resulting from deficiencies in the heuristics involved complicated coordination structures, which is not surprising, since coordination has always been problematic for TAG. In practice, we throw out all elementary tree templates seen only once in training, on the assumption that they are most likely the result of noise in the data or the extraction heuristics.

This method is extremely similar to that of Xia [132] and that of Chen [27], the main difference being that these other methods tend to add brackets to the Treebank

---

[2] This survey was performed on an earlier version of the extraction heuristics.

in order to obtain a more sensible grammar, whereas our method tends to reproduce the Treebank bracketing more closely, in order to facilitate comparison with other statistical parsers.

| | | ≤ 40 words | | | |
Model	LR	LP	CB	0CB	≤ 2 CB
PCFG [25]	71.7	75.8	2.03	39.5	68.1
Magerman [86]	84.6	84.9	1.26	56.6	81.4
Charniak [25]	87.5	87.4	1.0	62.1	86.1
present model	87.7	87.7	1.02	65.2	84.8
Collins [45]	88.5	88.7	0.92	66.7	87.1
Charniak [26]	90.1	90.1	0.74	70.1	89.6
		≤ 100 words			
Model	LR	LP	CB	0CB	≤ 2 CB
PCFG [25]	70.6	74.8	2.37	37.2	64.5
Magerman [86]	84.0	84.3	1.46	54.0	78.8
Charniak [25]	86.7	86.6	1.20	59.5	83.2
present model	87.0	87.0	1.21	62.2	82.2
Collins [45]	88.1	88.3	1.06	64.0	85.1
Charniak [26]	89.6	89.5	0.88	67.6	87.7

Table 3.1 Parsing results using heuristics on English. LR = labeled recall, LP = labeled precision; CB = average crossing brackets, 0 CB = no crossing brackets, ≤ 2 CB = two or fewer crossing brackets. All figures except CB are percentages.

We trained the model using the extraction heuristics on sections 02–21 and tested it on section 23. The results (Table 3.1) show that our parser lies roughly midway between the earliest [86] and latest [26] of the lexicalized PCFG parsers,[3] and that both do considerably better than a PCFG trained on Treebank trees *qua* CFG derivations [25]. While these results are not state-of-the-art, they demonstrate that a probabilistic TIG-SA parser can perform at the same level of accuracy as a lexicalized PCFG parser—or, under our reinterpretation, that lexicalization makes PCFG parsers perform at the same level of accuracy as a probabilistic TIG-SA parser.

We suspect that our model does not match the best of the lexicalized PCFG models because it is not using the larger elementary structures of TIG-SA very robustly. Fine-tuning the backoff model might bring accuracy closer to the state of the art, but it may be more productive to look to maximum-entropy models to provide greater flexibility in choosing model features with different amounts of context.

To see how this system would adapt to a different corpus in a different language, we replaced the head rules and argument/adjunct rules with rules appropriate for the Penn Chinese Treebank [133]. These were adapted from rules constructed by Xia [132] and are shown in Tables 3.6 and 3.7. We also retained the coordination rule described above.

We also made the following changes to the experimental setup:

---

[3] Note that these scores are an improvement over an earlier version [31].

- We lowered the unknown word threshold from five to one because the new training set was smaller.
- The POS tagger for unknown words had to be retrained on the new corpus.
- A beam width of 10^{-4} was used instead of 10^{-5}, for speed reasons.
- The comma pruning rule was not used, because it is based on an empirical observation from the English Treebank which does not hold for the Chinese Treebank.

We then trained the parser on sections 001–270 of the Penn Chinese Treebank (84,873 words) and tested it on sections 271–300 (6776 words). To provide a basis for comparison with performance on English, we performed two further tests. First, we trained the parser on sections 001–270 of the English translation of the Penn Chinese Treebank (118,099 words) and tested it on sections 271–300 (10,913) words). Second, we trained the parser on sections 02–03 of the WSJ corpus (82,592 words) and tested it on the first 400 sentences of section 23 (9473 words) with the same settings as the Chinese parser (but with the comma pruning rule). Note that because of the relatively small datasets used, cross-validation would be desirable for future studies.

Model	Corpus	≤ 100 words				
		LR	LP	CB	0CB	≤ 2 CB
present	WSJ-small	82.9	82.7	1.60	48.5	74.8
present	Xinhua-English	73.6	77.7	2.75	48.6	66.0
present	**Xinhua**	75.3	76.8	2.72	46.0	67.0
Model	Corpus	≤ 40 words				
		LR	LP	CB	0CB	≤ 2 CB
present	WSJ-small	83.5	83.1	1.42	50.4	77.2
present	Xinhua-English	76.4	82.3	1.39	61.4	78.3
present	**Xinhua**	78.4	80.0	1.79	52.8	74.8

Table 3.2 Parsing results using heuristics on Chinese. Abbreviations are as in Figure 3.1. Xinhua: trained on Penn Chinese Treebank sections 001–270, tested on sections 271–300. Xinhua-English: same, but on English translation. WSJ-small: trained on Penn Treebank, Wall Street Journal sections 02–03, tested on section 23, sentences 1–400.

The results, shown in Table 3.2,[4] show that this parser is quite usable on a language other than the one it was developed on. Indeed, it was the parser used to bootstrap later releases of the Penn Chinese Treebank, providing rough parses which human annotators can correct up to twice as fast as annotating from scratch [40].

---

[4] Because of part-of-speech tagging errors on the part of either the corpus or the parser, two sentences are flagged by the scorer as having the wrong length. The standard policy is to treat the sentence as if it were not in either the gold standard file or the test file, but the more rigorous policy used here is to keep the sentence but treat the test file as if it had not guessed any brackets. For the WSJ corpus, the scores are not affected, but in this case, they are affected slightly. Using the standard policy, labeled recall would be 75.8% for sentences of length ≤ 100 and 79.2% for sentences of length ≤ 40.

3.4.2 Training by Expectation-Maximization

In the type of approach we have been discussing so far, we use hand-written rules to reconstruct TIG-SA structural descriptions from the partial structural descriptions in training data, and then train the TIG-SA model by maximizing the likelihood of the reconstructed training data according to the model. However, the estimate we get will maximize the likelihood of the reconstructed training data, but not necessarily that of the observed training data itself. In this section we explore the possibility of training a model directly on partial structural descriptions using the method of Expectation-Maximization [47]; more specifically, a generalization of the Inside-Outside algorithm [79].

This approach follows a very similar experiment by Hwa [63]. The difference between the two is that whereas Hwa begins with a non-linguistically-motivated grammar that is designed to be general enough to generate any bracketing, we use the grammar and initial model induced by the heuristic method of the previous section. For this reason, Hwa's method is not able to take advantage of the information in the nonterminal labels in the data, but only the bracketing information.

The expectation step (E-step) of the Inside-Outside algorithm is performed by a parser that computes all possible derivations for each parse tree in the training data. This algorithm is analogous to CKY for TAG [125], except instead of items of the form $[\eta, i, j, k, l]$, where η ranges over elementary tree nodes and i, j, k, and l range over positions in the input string, it uses items of the form $[\eta, \theta_{il}, \theta_{jk}]$, where θ_{il} and θ_{jk} range over addresses in the input tree. By contrast, Hwa's E-step uses a standard TIG parser, but discards chart items with spans that cross a bracket in the input tree. But because the TIG parser achieves its cubic time complexity by pretending that foot nodes have zero spans, Hwa's implementation of the E-step will not work correctly if adjunction is allowed at spine nodes other than the root node. Since this type of adjunction is necessary to show that TIG has greater tree generative capacity than CFG, this implementation is not fully general.

The E-step then uses the derivation forest thus produced to compute inside and outside probabilities and uses these, in turn, to compute the expected number of times each decision occurred. Since a TIG-SA derivation forest has the same form as a CFG derivation forest, this computation is identical to the standard Inside-Outside algorithm; it is not necessary to define a specialized algorithm [112, 63].

For the maximization step (M-step), we obtain a maximum-likelihood estimate of the parameters of the model using relative-frequency estimation, just as in the original experiment, as if the expected values for the complete data were the training data; the only difference is that the expected values may be fractional.

Smoothing presents a special problem: recall that our model uses several backoff levels combined by linear interpolation. There are several ways one might incorporate this smoothing into the reestimation process, and we chose to depart as little as possible from the original smoothing method: in the E-step, we use the smoothed model, and after the M-step, we use the original formula (3.6) to recompute the smoothing weights based on the new counts computed from the E-step. While simple, this approach has two important consequences. First, since the formula for the smoothing weights intentionally does *not* maximize the likelihood of the training data, each iteration of reestimation is not guaranteed to increase the likelihood of the training data. Second, reestimation tends to increase the size of the model in memory, since smoothing gives nonzero expected counts to many choices which were unseen in training. Therefore, since the resulting model is quite large, if a choice at a particular point in the derivation forest has an expected count below 10^{-15}, we throw it out.

Another method would be to permanently use the smoothing weights computed on the initial model. This would restore the guarantee of nondecreasing likelihood, and perhaps limit the growth of the model as well. Bikel [14] has performed some initial experiments using this method.

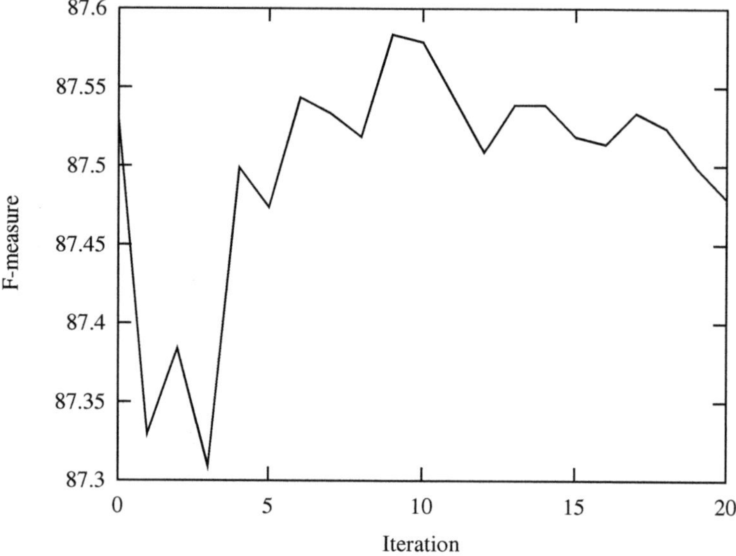

Fig. 3.7 Accuracy of reestimated models on held-out data: English, starting with full rule set.

We first trained the initial model on sections 02–21 of the WSJ corpus using the original head rules, and then ran the Inside-Outside algorithm on the same data. We tested each successive model on some held-out data (section 00), using a beam width of 10^{-4}, to determine at which iteration to stop. The F-measure (harmonic mean of labeled precision and recall) for sentences of length ≤ 100 for each iteration is shown in Figure 3.7. We then selected the ninth reestimated model and compared it with the initial model on section 23 (see Table 3.3). This model did only marginally better than the initial model on section 00, but it actually performs worse than the initial model on section 23. One explanation is that the head rules, since they have been extensively fine-tuned, do not leave much room for improvement. To test this, we ran two more experiments.

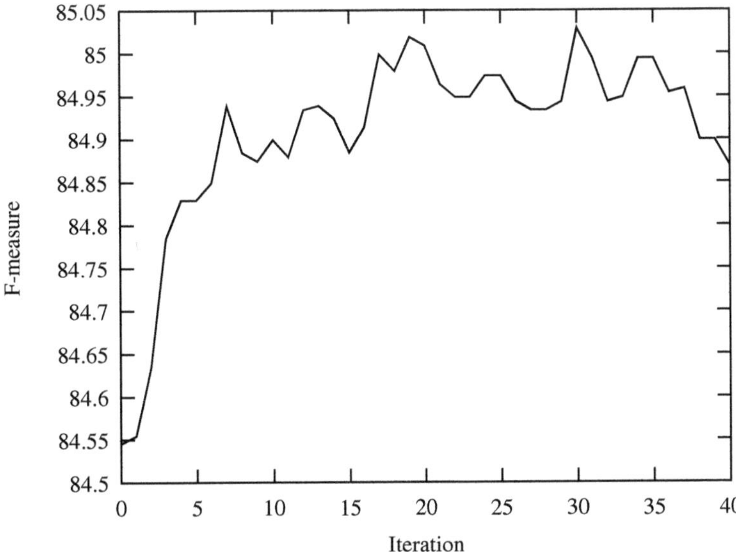

Fig. 3.8 Accuracy of reestimated models on held-out data: English, starting with simplified rule set.

The second experiment started with a simplified rule set, which simply chooses either the leftmost or rightmost child of each node as the head, depending on the label of the parent: e.g., for VP, the leftmost child is chosen; for NP, the rightmost child is chosen. The argument rules, however, were not changed. This rule set is supposed to represent the kind of rule set that someone with basic familiarity with English syntax might write down in a few minutes. The reestimated models seemed to improve on this simplified rule set when parsing section 00 (see Figure 3.8); however, when we compared the 30th reestimated model with the initial model on section 23 (see Figure 3.3), there was no improvement.

The third experiment was on the Chinese Treebank, starting with the same head rules used in [15]. These rules were originally written by Xia for grammar development, and although we have modified them for parsing, they have not received as much fine-tuning as the English rules have. We trained the model on sections 001–270 of the Penn Chinese Treebank, and reestimated it on the same data, testing it at each iteration on sections 301–325 (Figure 3.9). We selected the 38th reestimated model for comparison with the initial model on sections 271–300 (Figure 3.3). Here we did observe a small improvement: an error reduction of 3.4% in the F-measure for sentences of length ≤ 40.

Our hypothesis that reestimation does not improve on the original rule set for English because that rule set is already fine-tuned was partially borne out by the second and third experiments. The model trained with a simplified rule set for English showed improvement on held-out data during reestimation, but showed no improvement in the final evaluation; however, the model trained on Chinese did show

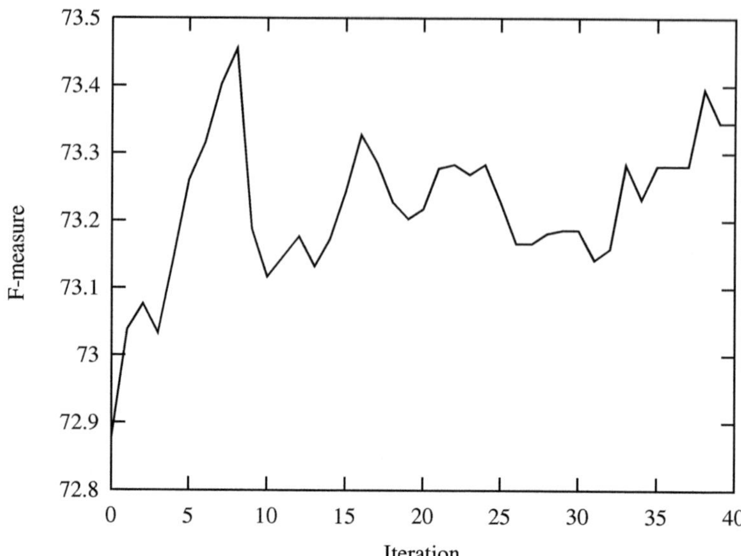

Fig. 3.9 Accuracy of reestimated models on held-out data: Chinese, starting with full rule set.

		≤ 100 words				
Model/Corpus	Step	LR	LP	CB	0CB	≤ 2 CB
English	initial	87.0	87.0	1.21	62.4	82.3
	9	86.4	86.7	1.26	61.4	81.8
English-simple	initial	84.5	84.2	1.54	57.6	78.4
	30	84.2	84.5	1.53	58.0	77.8
Chinese	initial	75.3	76.8	2.72	46.0	67.0
	38	75.2	78.0	2.66	47.7	67.6
		≤ 40 words				
Model/Corpus	Step	LR	LP	CB	0CB	≤ 2 CB
English	initial	87.7	87.8	1.02	65.3	84.9
	9	87.2	87.5	1.06	64.4	84.2
English-simple	initial	85.5	85.2	1.29	60.7	81.1
	30	85.1	85.4	1.30	60.9	80.6
Chinese	initial	78.4	80.0	1.79	52.8	74.8
	38	78.8	81.1	1.69	54.2	75.1

Table 3.3 Parsing results using EM. Original = trained on English with original rule set; Simple = English, simplified rule set. LR = labeled recall, LP = labeled precision; CB = average crossing brackets, 0 CB = no crossing brackets, ≤ 2 CB = two or fewer crossing brackets. All figures except CB are percentages.

a small improvement in both. We are uncertain as to why the gains observed during the second experiment were not reflected in the final evaluation, but based on the graph of Figure 3.8 and the results on Chinese, we believe that reestimation by EM can be used to facilitate adaptation of parsing models to new languages or corpora.

3.5 Related Work

Subsequent to the work described in this chapter, there have been more ambitious and successful attempts to learn probabilistic CFGs from partial structural descriptions [100, 88], culminating in the Berkeley parser [98]. In these models, the nonterminal symbols of the CFG are the nonterminal symbols observed in the training data, augmented with a small set of hidden states. No assumptions are made about the meaning of the states; the rule probabilities are simply randomly initalized and estimated using Expectation-Maximization. When this is done carefully, the resulting probabilistic CFG achieves state-of-the-art performance across a number of languages.

In one respect, this work is a confirmation of the theoretical framework we have presented here; in another respect, it is a conundrum. It is a confirmation of the central prediction of the framework, namely, formalisms that extend CFG while still being coverable by CFG (respecting trees) have the same SGC as CFG with respect to the domain of weighted trees; therefore, they should not be able to describe statistical parsing models that are any better than probabilistic CFG. In particular, probabilistic TSG, RF-TAG, and component-local multicomponent CFG should not be any better than probabilistic CFG. The Berkeley parser [98] confirms that excellent parsing performance can indeed be achieved using probabilistic CFG.

On the other hand, because this CFG is constructed automatically, there seems to be little hope of reversing the covering construction as we have done here for the Collins parser [45]. That is, there is no clear way to reimagine the Berkeley parser as the cover grammar or an approximation of a cover grammar for some grammar formalism that generates trees directly (without annotations like heads or states) and to reconstruct the original grammar formalism. The hidden states of the Berkeley parser are claimed to be interpretable, and the states attached to part-of-speech tags indeed have clear meanings (for example, NNP clearly splits into subcategories for surnames, middle initials, place names, names of months or days of the week, etc.), but it is much harder to see how the states attached to nonterminal symbols pass useful information through the tree. Perhaps a different formalism could still shed some light on why the Berkeley parser performs so well.

3.6 Summary

We have shown how to define parsing models of a general nature based on grammars, and found that many of the grammar formalisms that have been proposed to increase the SGC of CFG in fact do not define any more parsing models than PCFG does. We then showed how a lexicalized PCFG can be thought of as a compiled version of a model defined over richer structural descriptions than are found in typical treebanks, and described our implementation of this new view in a probabilistic TIG-SA which performs at the same level of accuracy as lexicalized PCFG. This result demonstrates that TIG-SA is a viable framework for statistical parsing. Moreover, it provides more flexibility than the head- and argument-finding rules of current lexicalized PCFG models.

Parent	Child
ADJP	first NNS; first QP; first NN; first \$; first ADVP; first JJ; first VBN; first VBG; first ADJP; first JJR; first NP; first JJS; first DT; first FW; first RBR; first RBS; first SBAR; first RB; first *
ADVP	last RB; last RBR; last RBS; last FW; last ADVP; last TO; last CD; last JJR; last JJ; last IN; last NP; last JJS; last NN; last *
CONJP	last CC; last RB; last IN; last *
FRAG	last *
INTJ	first *
LST	last LS; last :; last *
NAC	first NN; first NNS; first NNP; first NNPS; first NP; first NAC; first EX; first \$; first CD; first QP; first PRP; first VBG; first JJ; first JJS; first JJR; first ADJP; first FW; first *
{NP, NX}	last {NN, NNP, NNPS, NNS, NX, POS, JJR}
	first NP
	last {\$, ADJP, PRN}; last CD; last {JJ, JJS, RB, QP}; last *
PP	last IN; last TO; last VBG; last VBN; last RP; last FW; last *
PRN	first *
PRT	last RP; last *
QP	first \$; first IN; first NNS; first NN; first JJ; first RB; first DT; first CD; first NCD; first QP; first JJR; first JJS; first *
RRC	last VP; last NP; last ADVP; last ADJP; last PP; last *
S	first TO; first IN; first VP; first S; first SBAR; first ADJP; first UCP; first NP; first *
SBAR	first WHNP; first WHPP; first WHADVP; first WHADJP; first IN; first DT; first S; first SQ; first SINV; first SBAR; first FRAG; first *
SBARQ	first SQ; first S; first SINV; first SBARQ; first FRAG; first *
SINV	first VBZ; first VBD; first VBP; first VB; first MD; first VP; first S; first SINV; first ADJP; first NP; first *
SQ	first VBZ; first VBD; first VBP; first VB; first MD; first VP; first SQ; first *
UCP	last *
VP	first TO; first VBD; first VBN; first MD; first VBZ; first VB; first VBG; first VBP; first VP; first ADJP; first NN; first NNS; first NP; first *
WHADJP	first CC; first WRB; first JJ; first ADJP; first *
WHADVP	last CC; last WRB; first *
WHNP	first WDT; first WP; first WP\$; first WHADJP; first WHPP; first WHNP; first *
WHPP	last IN; last TO; last FW; last *
X	first *

Table 3.4 Head rules for the Penn (English) Treebank. Rules (delimited by line breaks or semi-colons) apply sequentially for each parent node until a match is found. The symbol * stands for any label.

Parent	Child
S	all {NP, SBAR, S} except A
{VP, SQ, SINV}	all {NP, SBAR, S, VP} except A
SBAR	all S except A
SBARQ	all SQ except A
NP	all NP except A
PP	first {PP, NP, WHNP, ADJP, ADVP, S, SBAR, VP, UCP} after head

where $A = \{$-ADV, -VOC, -BNF, -DIR, -EXT, -LOC, -MNR, -TMP, -PRP, -CLR$\}$

Table 3.5 Argument rules for the Penn (English) Treebank. All rules apply to every parent-child pair.

Parent	Child
ADJP	last {ADJP, JJ}; last {AD, NN, CS}; last *
ADVP	last {ADVP, AD}; last *
CLP	last {CLP, M}; last *
CP	last {DEC, SP}; first {ADVP, CS}; last {CP, IP}; last *
DNP	last {DNP, DEG}; last DEC; last *
DVP	last {DVP, DEV}; last *
DP	first {DP, DT}; first *
FRAG	last {VV, NR, NN}; last *
INTJ	last {INTJ, IJ}; last *
LST	first {LST, CD, OD}; first *
IP	last {IP, VP}; last VV; last *
LCP	last {LCP, LC}; last *
NP	last {NP, NN, NT, NR, QP}; last *
PP	first {PP, P}; first *
PRN	last {NP, IP, VP, NT, NR, NN}; last *
QP	last {QP, CLP, CD, OD}; last *
VP	first {VP, VA, VC, VE, VV, BA, LB, VCD, VSB, VRD, VNV, VCP}; first *
VCD	last {VCD, VV, VA, VC, VE}; last *
VRD	last {VRD, VV, VA, VC, VE}; last *
VSB	last {VSB, VV, VA, VC, VE}; last *
VCP	last {VCP, VV, VA, VC, VE}; last *
VNV	last {VNV, VV, VA, VC, VE}; last *
VPT	last {VPT, VV, VA, VC, VE}; last *
UCP	last *
WHNP	last {WHNP, NP, NN, NT, NR, QP}; last *
WHPP	first {WHPP, PP, P}; first *

Table 3.6 Head rules for the Penn Chinese Treebank. Rules (delimited by line breaks or semi-colons) apply sequentially for each parent node until a match is found. The symbol * stands for any label.

Parent	Child
VP	all {CP, IP, VP} except -ADV
CP	all {CP, IP} except -ADV
PP	all {NP, DP, QP, LCP, CP, IP, UCP} except -ADV
DNP	all {NP, DP, QP, LCP, PP, ADJP, UCP} except -ADV
DVP	all {NP, DP, QP, VP, ADVP, UCP} except -ADV
LCP	all {NP, DP, QP, LCP, IP, PP, UCP} except -ADV
*	all {-SBJ, -OBJ, -IO, -PRD} except -ADV

Table 3.7 Argument rules for the Penn Chinese Treebank. All rules apply to every parent-child pair. The symbol * stands for any label.

Chapter 4
Machine Translation

In this chapter we discuss applications of grammars to natural language translation. We define how to measure the power of grammar formalisms for translation and find that this domain classifies formalisms more finely than in the previous chapter: formalisms which were previously equivalent now separate into different levels of power. This means that machine translation stands to gain more from richer grammar formalisms than statistical parsing does; on the other hand, it is all the more important for machine translation research to find the right level of power and not give up on grammar-based methods. We discuss the formal properties of synchronous RF-TAG and the possibility of its use for language translation.

4.1 Measuring Translation Power

We can measure the translation power of a grammar formalism by its SGC with respect to the domain of string pairs.

Definition 34. The interpretation domain of *string pairs* has interpretations which are pairs of (tuples of) strings. For a formalism $\mathscr{F}$, let Φ_{string} be the interpretation function in the domain of strings. Then Φ, the interpretation function in the domain of string pairs, is

$$\Phi(G) = \{f_s \times f_t \mid \Phi_{\text{string}}(G) = \{f_s\}, \Phi_{\text{string}}(G') = \{f_t\}, \forall G' \in \mathscr{F}\} \tag{4.1}$$

In other words, the interpretation of a derivation τ is the string yield of τ and the string yield of τ in some other grammar from $\mathscr{F}$.

For example, the following synchronous sLMG recognizes a simple English sentence, and its interpretations in the domain of string pairs are translations into Japanese, which has SOV word order:

$$S(xy) :- NP(x), VP(y) \qquad \langle xy, x'y' \rangle \hookleftarrow \langle \langle x, x' \rangle, \langle y, y' \rangle \rangle \qquad (4.2)$$
$$VP(xy) :- V(x), NP(y) \qquad \langle xy, y'x' \rangle \hookleftarrow \langle \langle x, x' \rangle, \langle y, y' \rangle \rangle \qquad (4.3)$$
$$NP(i) \qquad \langle i, watashi\ wa \rangle \qquad (4.4)$$
$$NP(the\ box) \qquad \langle the\ box, hako\ wo \rangle \qquad (4.5)$$
$$V(open) \qquad \langle open, akemasu \rangle \qquad (4.6)$$

However, it is more common to write grammars for translation as *synchronous grammars*, which generate pairs of strings or structures in a way that is symmetric between the source and target. The synchronous grammar formalisms we consider here are subclasses of *synchronous linear sLMG*, introduced by Bertsch and Nederhof [12] as *simple range concatenation transducers*. We could redefine synchronous sLMGs as follows:

Definition 35. A *synchronous production* is a multicomponent linear sLMG production (see Section 2.5.2) whose predicates have exactly two components, called the *source* and *target* components. All the arguments from the source component of each right-hand side predicate appear only in the source component of the lefthand side, and all the arguments from the target component of each right-hand side predicate appear only in the target component of the left-hand side.

Definition 36. A *synchronous sLMG* is a linear sLMG whose productions are all synchronous productions and with a two-component start predicate S. The string relation defined by a synchronous linear sLMG is the set

$$\{ \langle w_1, w_2 \rangle \mid S(w_1 : w_2) \text{ is derivable} \}$$

The above example as a synchronous sLMG would read:

$$S(xy : x'y') :- NP(x : x'), VP(y : y') \qquad (4.7)$$
$$VP(xy : y'x') :- V(x : x'), NP(y : y') \qquad (4.8)$$
$$NP(i : watashi\ wa) \qquad (4.9)$$
$$NP(the\ box : hako\ wo) \qquad (4.10)$$
$$V(open : akemasu) \qquad (4.11)$$

Definition 37. The *source projection* of a synchronous sLMG (according to Definition 36) is the sLMG formed by removing all the target components from all the arguments in the grammar. Similarly, the *target projection* is the sLMG formed by removing all the source components.

The reason for the linearity requirement is to guarantee that the projections of a synchronous sLMG have the same derivation sets as the original.

It is not hard to see that the string-pair yields of a synchronous sLMG can also be seen as interpretations of its source projection. Therefore, the WGC of synchronous $\mathscr{F}$ (that is, the set of string pairs it generates) is the same as the SGC of $\mathscr{F}$ with respect to the domain of string pairs. This is the measure we use for the translation power of a grammar formalism. If a grammar formalism can generate more string relations, it will be a more flexible system for defining translations. Tree relations are also a useful measure when constraints on phrase structures are needed.

4.2 Translation and Bitext Parsing

The two most common algorithms used with synchronous grammars are *translation* and *bitext parsing*. In translation, we are given a source string and must compute the possible target strings it translates to. To do this, we simply project the synchronous grammar to its source side, apply the parsing algorithm of Section 2.5.3, and then map from derivations to target strings. Thus translation is no more difficult than parsing.[1]

In bitext parsing, the input is a pair of input strings $w : w'$ in the source and target language, and our goal is to compute its possible derivations under the synchronous grammar G. This is needed, for example, when using the Inside-Outside algorithm to estimate production probabilities from a parallel text. To do this, we take G (written as a synchronous sLMG) and add a single top-level production:

$$S'(x\$x') :- S(x : x') \tag{4.12}$$

where S' and $\$$ are a special nonterminal and terminal symbol, respectively, not found elsewhere in G, and use the parsing algorithm of Section 2.5.3 to parse the string $w\$w'$.

Usually, the computational complexity of bitext parsing is the square of that of translation. Thus, since parsing with a CFG in Chomsky normal form is $\mathscr{O}(n^3)$, bitext parsing with a synchronous CFG in Chomsky normal form is $\mathscr{O}(n^6)$. However, as we will see below, synchronous grammars often have different properties from their non-synchronous counterparts. For example, whereas a CFG can always be converted into Chomsky normal form, a synchronous CFG in general cannot.

4.3 Synchronous CFG

Synchronous CFGs were originally known as *syntax-directed transduction grammars* [81] or *syntax-directed translation schemata* [4]. We saw an example of a synchronous CFG above; in more standard notation, this grammar would be:

---

[1] However, when intersecting the target side grammar with an n-gram language model, as is common in machine translation [130, 36], the computational complexity of translation becomes more like that of bitext parsing.

$$\langle S \to NP_{\boxed{1}}\ VP_{\boxed{2}}, S \to NP_{\boxed{1}}\ VP_{\boxed{2}} \rangle \tag{4.13}$$

$$\langle VP \to V_{\boxed{1}}\ NP_{\boxed{2}}, VP \to NP_{\boxed{2}}\ V_{\boxed{1}} \rangle \tag{4.14}$$

$$\langle NP \to i, NP \to watashi\ wa \rangle \tag{4.15}$$

$$\langle NP \to the\ box, NP \to hako\ wo \rangle \tag{4.16}$$

$$\langle V \to open, V \to akemasu \rangle \tag{4.17}$$

The boxed numbers $\boxed{i}$ link up nonterminal symbols on the source side with nonterminal symbols on the target side: $\boxed{1}$ links to $\boxed{1}$, $\boxed{2}$ with $\boxed{2}$, and so on.

Just as we start in a CFG with a start symbol and repeatedly rewrite nonterminal symbols using the productions, so in a synchronous CFG, we start with a pair of linked start symbols (with the starting index 10 chosen arbitrarily),

$$\langle S_{\boxed{10}}, S_{\boxed{10}} \rangle$$

and repeatedly rewrite pairs of nonterminal symbols using the productions—with two wrinkles. First, when we apply a production, we renumber the boxed indices consistently to fresh indices that aren't in our working string pair. Thus, applying production (4.13), we get

$$\Rightarrow \langle NP_{\boxed{11}}\ VP_{\boxed{12}}, NP_{\boxed{11}}\ VP_{\boxed{12}} \rangle$$

Second, we are only allowed to rewrite *linked* nonterminal symbols. Thus we can apply production (4.14) like so:

$$\Rightarrow \langle NP_{\boxed{11}}\ V_{\boxed{13}}\ NP_{\boxed{14}}, NP_{\boxed{11}}\ NP_{\boxed{14}}\ V_{\boxed{13}} \rangle$$

But now if we want to apply production (4.15), we can't apply it to $NP_{\boxed{11}}$ on one side and $NP_{\boxed{14}}$ on the other, like this:

$$\not\Rightarrow \langle i\ V_{\boxed{13}}\ NP_{\boxed{14}}, NP_{\boxed{11}}\ watashi\ wa\ V_{\boxed{13}} \rangle$$

But we can apply it to any linked nonterminals, like so:

$$\Rightarrow \langle i\ V_{\boxed{13}}\ NP_{\boxed{14}}, watashi\ wa\ NP_{\boxed{14}}\ V_{\boxed{13}} \rangle$$

$$\Rightarrow \langle i\ open\ NP_{\boxed{14}}, watashi\ wa\ NP_{\boxed{14}}\ akemasu \rangle$$

$$\Rightarrow \langle i\ open\ the\ box, watashi\ wa\ hako\ wo\ akemasu \rangle$$

Thus we have an English string and Japanese string which are translations of each other.

4.3.1 Applications to Translation

Inversion transduction grammar

Define the *rank* of a production to be the number of nonterminals on its right-hand side (for CFG) or in the source or target component of its right-hand side (for synchronous CFG). Whereas any CFG can be converted into a weakly equivalent CFG whose productions have rank at most two, there exists, for all $r \geq 4$, a synchronous CFG of rank r that cannot be converted into an equivalent (with respect to the domain of string pairs) grammar of rank $(r-1)$ [4]. For example, the following rank-4 production cannot be reduced to lower-rank productions:

$$\langle A \rightarrow B\boxed{1}\ C\boxed{2}\ D\boxed{3}\ E\boxed{4}, A \rightarrow D\boxed{3}\ B\boxed{1}\ E\boxed{4}\ C\boxed{2}\rangle \tag{4.18}$$

The case $r = 2$ is commonly known as *inversion transduction grammar* [131]. Inversion transduction grammars were proposed by Wu [131] as a candidate formalism for statistical machine translation. He proposed efficient algorithms for this purpose, including for decoding with an n-gram target-side language model [130].

Yamada-Knight syntax-based model

The translation model of Yamada and Knight [134, 135] is formally a synchronous CFG in which French productions are generated from English productions (following the convention of Brown et al. [24]) through a sequence of transformations.

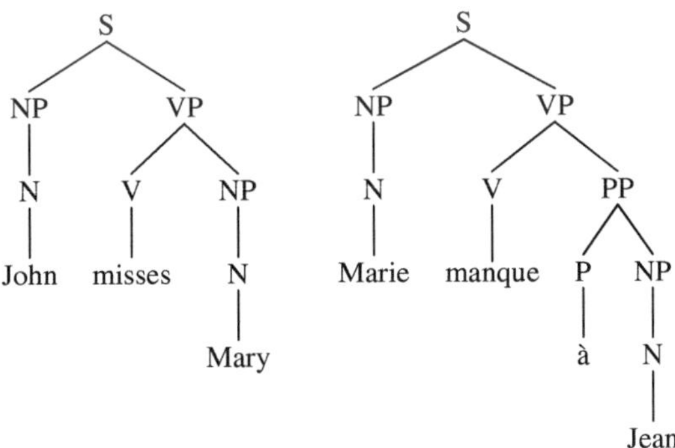

Fig. 4.1 Example of a French-English tree pair that cannot be generated by synchronous CFG.

Above, we considered synchronous CFGs as defining relations on strings. Thought of as defining a relation on trees, synchronous CFGs can only relabel nodes and reorder sisters. Thus we cannot write a synchronous CFG that can swap subjects and objects, as in

(4.19)　John misses Mary.

(4.20)　Marie manque à Jean.
　　　　　Mary misses to John

with trees shown in Figure 4.1, because $[_{\text{NP}}$ John$]$ and $[_{\text{NP}}$ Mary$]$ are not sister nodes and cannot be swapped.

However, if we are willing to alter the trees, then there is an easy solution:

$$S \rightarrow \langle \text{NP}① \text{ misses NP}②, \text{NP}② \text{ manque à NP}① \rangle \tag{4.21}$$

$$\text{NP} \rightarrow \langle \text{John}, \text{John} \rangle \tag{4.22}$$

$$\text{NP} \rightarrow \langle \text{Mary}, \text{Mary} \rangle \tag{4.23}$$

By flattening the trees, we made $[_{\text{NP}}$ John$]$ and $[_{\text{NP}}$ Mary$]$ sister nodes so that they could be swapped. (We also integrated the verb *misses/manque à* into the rule to make sure that this swapping only occurs with this particular verb.)

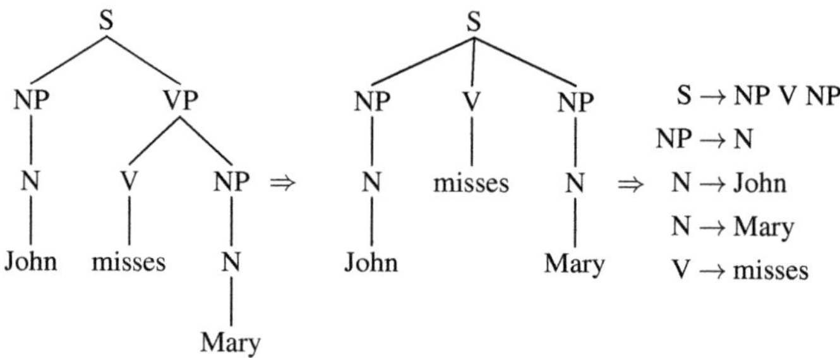

Fig. 4.2 Flattening of trees in the translation system of Yamada and Knight [134].

Yamada and Knight's system flattens English trees by deleting nodes that have the same head word as their parent according to the Magerman rules (see Figure 4.2). In particular, it deletes the VP nodes in the above example, enabling it to perform subject-object swapping.

Hiero

Phrase-based translation models [95, 74] dramatically advanced the state of the art by learning translations of multi-word units. The Hiero translation model [35, 36] combines the formalism of synchronous CFG with phrase-based translation's ability to learn multi-word units. The synchronous CFG used in Hiero has a nonterminal alphabet with only two symbols, $\{S, X\}$. The grammar consists of two kinds of rules. Most rules are of the form

$$\langle X \to \gamma, X \to \alpha \rangle \tag{4.24}$$

where γ consists of French terminals and at most two occurrences of the nonterminal X, and α consists of English terminals and at most two occurrences of the nonterminal X. These rules can be automatically extracted from word-aligned parallel text. The resulting grammar is large: a training corpus with hundreds of millions of words can lead to a grammar with hundreds of millions of rules. In addition to these rules, the grammar contains the two so-called *glue rules*

$$\langle S \to S_{\boxed{1}}\, X_{\boxed{2}}, S \to S_{\boxed{1}}\, X_{\boxed{2}} \rangle \tag{4.25}$$

$$\langle S \to X_{\boxed{1}}, S \to X_{\boxed{1}} \rangle \tag{4.26}$$

These allow the grammar to break a sentence into a sequence of smaller chunks; since reordering does not occur among the chunks, this speeds up translation. Finally, the decoder dynamically intersects the target side of this synchronous grammar with an n-gram language model. The resulting translation system can improve significantly over phrase-based translation models, particularly on Chinese-English translation.

4.3.2 Extensions

Synchronous TSG

Synchronous tree-substitution grammars [111, 51] are able to perform subject-object swapping as in Figure 4.1 without flattening the trees. In a synchronous TSG, the productions are pairs of elementary trees, and the leaf nonterminals are linked just as in synchronous CFG (see Figure 4.3). Thus, if we care about the trees produced, synchronous TSG is a more powerful alternative to synchronous CFG.

Proposition 4. *Synchronous TSG is strongly equivalent to synchronous CFG with respect to the domain of string pairs but has greater SGC with respect to the domain of tree pairs.*

Proof. Any synchronous TSG can easily be converted into a synchronous CFG generating the same string pairs: for each tree pair $\langle \alpha_1 : \alpha_2 \rangle$, create the production pair

$$\langle root(\alpha_1) \to yield(\alpha_1) : root(\alpha_2) \to yield(\alpha_2) \rangle$$

where *root* and *yield* are functions returning the root and yield, respectively, of a tree.

On the other hand, the following synchronous TSG generates a trivial example of a tree relation that no synchronous CFG can generate:

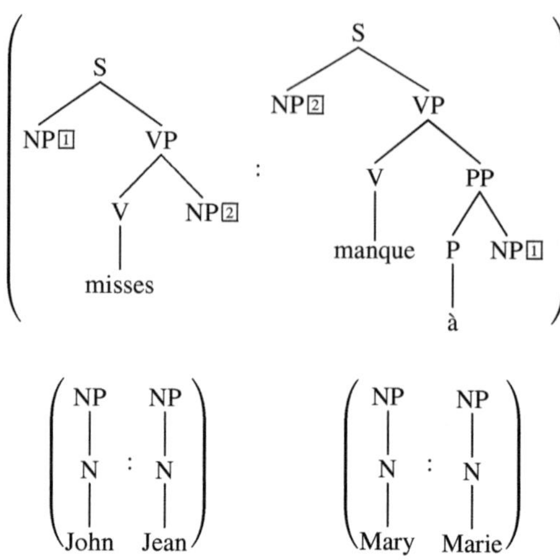

Fig. 4.3 Synchronous TSG demonstrating transposition of subject and object.

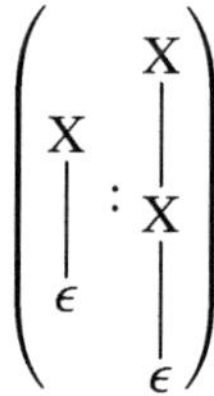

because the tree relations generated by a synchronous CFG must always have equal numbers of nonterminal nodes.

Synchronous TSG was proposed by Eisner [51] and Gildea [55] for use in statistical machine translation. However, in a certain sense this had already been achieved by the model of Yamada and Knight [134], which performed subject-object swapping within a synchronous CFG by altering the trees. Just as the tree transformations in Collins' parser could be reinterpreted as a cover grammar for a TIG-SA, we can also reinterpret Yamada and Knight's model as a cover grammar for a synchronous TSG (see Figure 4.4).

The string-to-tree model of Galley et al. [53, 52] and the tree-to-string models of Liu et al. [82] and Huang et al. [60] are essentially synchronous TSGs, where the source or target side, respectively, consists only of single-level elementary trees. These grammars are extracted from parallel data where only one side has been parsed; full elementary trees are extracted from the parsed side, whereas single-level trees are extracted from the unparsed side, as in Hiero. These systems have been quite successful, generally outperforming Hiero.

However, implementations of full synchronous TSG, that is, systems that generate linguistically reasonable trees on both sides, turn out to be overconstrained by the syntax of the two languages [8, 83]. One solution is to restructure the trees [8];

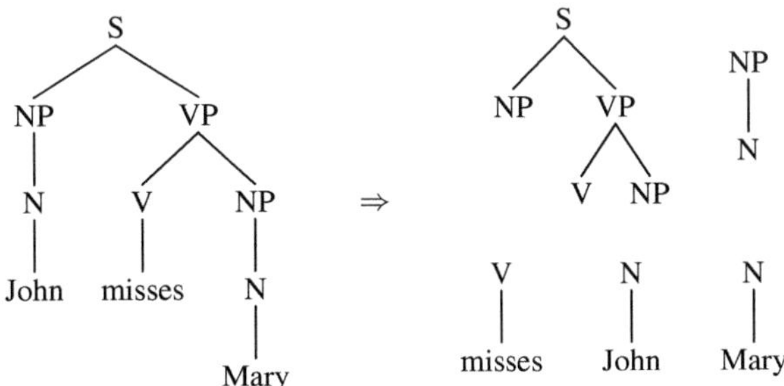

Fig. 4.4 Alternative view of flattening operation as extraction of unlexicalized TSG.

another is to extract grammars from forests of possible parses instead of a singly, possibly noisy, parse [83]. A third solution is to extend the formalism beyond synchronous TSG.

Synchronous tree sequence substitution grammar

For example, Zhang et al. [136] move to *synchronous tree sequence substitution grammar*, which could be thought of as restriction of multicomponent TSG in which the components of a set must substitute into sister substitution sites. This is a very mild extension of TSG, but already seems to alleviate some of the most serious nonisomorphisms that pure synchronous TSG cannot handle.

Synchronous TIG

Synchronous TIG extends synchronous TSG with an adjunction operation. Since TIG has greater tree generative capacity than CFG, it follows that synchronous TIG can generate more tree relations than synchronous CFG. However, the two formalisms generate exactly the same string relations, since the conversion from TIG to CFG given by Schabes and Waters [116] preserves derivations [33].

Nesson et al. [92] have experimented with estimating the parameters of a small probabilistic synchronous TIG from parallel data. DeNeefe and Knight [48] have carried out much larger-scale experiments on a generalization of the Galley et al. [53, 52] system to synchronous TIG, with promising results.

4.4 Synchronous TAG

Synchronous TAG was first defined by Shieber [123] as a correction of an older faulty definition [124]. The new definition, which requires that the source and target derivations be isomorphic, is essentially a generalization to TAG of synchronous CFG. In this definition, a synchronous TAG is a set of pairs of elementary trees in which input interior/substitution nodes are coindexed with output interior/substitution nodes.

Consider again the case of Dutch cross-serial dependencies from Section 1.3:

(4.27)   (that) John saw Peter help the children swim

(4.28)   (dat) Jan   Piet   de   kinderen zag   helpen zwemmen
         (that) John Peter the children saw   help   swim

If we assume that both languages can stack an unbounded number of clauses this way, then this translation between Dutch and English is provably beyond the power of synchronous CFG and TSG. But it is easy to pair the TAG elementary trees from Figure 1.1 with English elementary trees to peform this translation.

4.4.1 Synchronous Regular Form TAG

Synchronous RF-TAG has been proposed by Dras [49] as a meta-level grammar for generating pairs of TAG derivation trees (see Section 4.4.2 below). We may also consider synchronous RF-TAG on its own. Although weakly context-free, it has strictly greater SGC with respect to the domain of string pairs (and therefore tree pairs) than synchronous CFG does [33].

Lemma 1 (synchronous pumping lemma). *Let L be a string relation generated by a synchronous CFG. Then there is a constant n such that if $\langle z : z' \rangle$ is in L and $|z| \geq n$ and $|z'| \geq n$, then $\langle z : z' \rangle$ may be written as $\langle uwy : u'w'y' \rangle$, and there exist strings v, x, v', x', such that $|vxv'x'| > 0$, $|vwx| \leq n$, $|v'w'x'| \leq n$, and for all $i \geq 0$, $\langle uv^i wx^i y : u'v'^i w'x'^i y' \rangle$ is in L.*

Proof. The proof is analogous to that of the standard pumping lemma [59, pp. 125–127]. However, a CFG cannot be put into Chomsky normal form without changing its SGC with respect to the domain of string pairs, and there are both sides to take into account. So we let m be the longest right-hand side in either grammar or the number 2, whichever is greater, and k and k' be the size of the two nonterminal alphabets; then $n = m^{kk'}$. This guarantees the existence of a pair of corresponding paths in the derivation of $\langle z : z' \rangle$ such that the same pair of nonterminals $\langle A : A' \rangle$ occurs twice:

$$\langle S : S \rangle \overset{*}{\Rightarrow} \langle uAy : u'A'y' \rangle \overset{*}{\Rightarrow} \langle u_1 vAxy_1 : u_1' v'A'x'y_1' \rangle \overset{*}{\Rightarrow} \langle u_1 vwxy_1 : u_1' v'w'x'y_1' \rangle$$

If we let $u = u_1 v$ and $y = x y_1$, and likewise $u' = u'_1 v'$ and $y' = x' y'_1$, then $\langle z : z' \rangle = \langle uwy : u'w'y' \rangle$, and for all $i \geq 0$, $\langle uv^i wx^i y : u'v'^i w'x'^i y' \rangle \in L$.

Proposition 5. *The string relation*

$$L = \{\langle a^m b^n c^n d^m : b^n a^m d^m c^n \rangle\}$$

is generable by a synchronous RF-TAG but not by any synchronous CFG.

Proof. The following synchronous RF-TAG generates L:

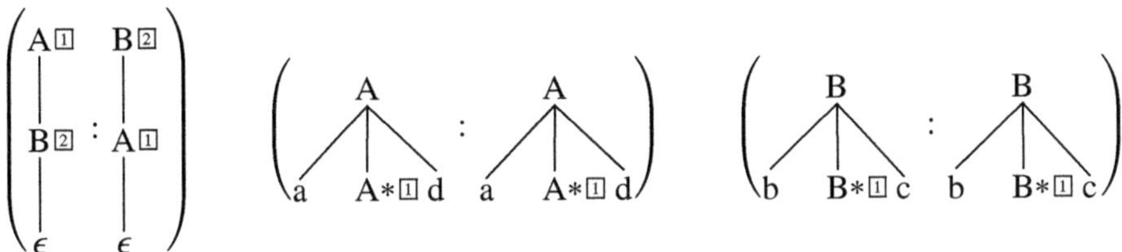

But suppose L can be generated by some CFG G. For any n given by the pumping lemma, let $\langle z : z' \rangle = \langle a^n b^n c^n d^n : b^n a^n d^n c^n \rangle$ satisfy the conditions of the pumping lemma. Then $vxv'x'$ must contain only a's and d's, or only b's and c's, otherwise $\langle uv^i wx^i y : u'v'^i w'x'^i y' \rangle$ will not be in L. But in the former case, $|vwx| > n$, and in the latter case, $|v'w'x'| > n$, which is a contradiction.

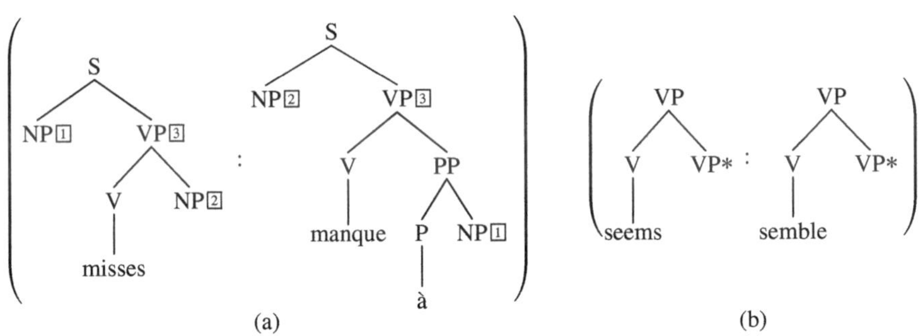

Fig. 4.5 Synchronous TAG fragment demonstrating long-distance transposition of subject and object.

We saw above how synchronous TSG can be used to transpose subjects and objects. But synchronous TSG cannot perform the same translation with a raising verb added:

(4.29)   John seems to miss Mary.

(4.30)   Marie semble manquer à Jean.
         Mary seems to miss   to John

To do that, we need adjunction of an auxiliary tree pair (Figure 4.5), which stretches the subject-object transposition apart arbitrarily. This is within the power of both TIG and RF-TAG.[2]

4.4.2 Extensions

We can extend synchronous TAG by analogy with the above extensions of synchronous CFG, using the notion of a *meta-level grammar*, also known as a *control grammar* [129, 49]. In this arrangement, a grammar G_1 (the meta-level grammar) generates the derivations of another grammar G_2 (the object-level grammar). Together they can be thought of as a single grammar, $G_2 \circ G_1$. In the examples below, the object-level grammar will be a TAG, and the meta-level grammar will generate recognizable sets of trees (that is, tree sets that can be generated by a CFG and a projection of labels).

Bounded nonisomorphisms: TAG ∘ TSG

Shieber notes that synchronous TAG (under the corrected definition) is too rigid to handle the translation of certain constructions. He suggests that the isomorphism requirement be relaxed to allow "bounded nonisomorphisms." He does not define this extension formally, but we may retroject Dras's use of control or meta-level grammars [49] into Shieber's suggestion, defining it as synchronous TAG ∘ TSG. This system adds some flexibility to synchronous TAG, but does not permit any more string-to-string mappings to be described.

Proposition 6. *Synchronous TAG ∘ TSG is strongly equivalent to synchronous TAG with respect to the domain of string pairs.*

As an example of how a meta-level grammar works and how synchronous TAG ∘ TSG can be useful for translation, consider the nonisomorphism which Shieber calls "elimination of dominance" in the following sentences:

(4.31)   The doctor treats his teeth.

(4.32)   Le docteur lui   soigne les dents.
         the doctor   him treats   the teeth

Since 'his' would adjoin into the tree for 'teeth' in (4.31), but 'lui' would not be able to adjoin into the tree for 'les dents' in (4.32) but have to adjoin into the tree for

---

[2] In our definition of RF-TAG, adjunction is allowed at foot nodes but not auxiliary root nodes, contrary to typical TAG syntactic analyses. We could accept the divergence, or else we could change the definition of RF-TAG to allow adjunction at root nodes instead of foot nodes, but this would make parsing less simple.

'soigne' instead, the two derivations will not be isomorphic (see Figure 4.7). This nonisomorphism is bounded, however, because 'his'/'lui' will never be more than two steps away in the derivation tree from the object it semantically modifies.

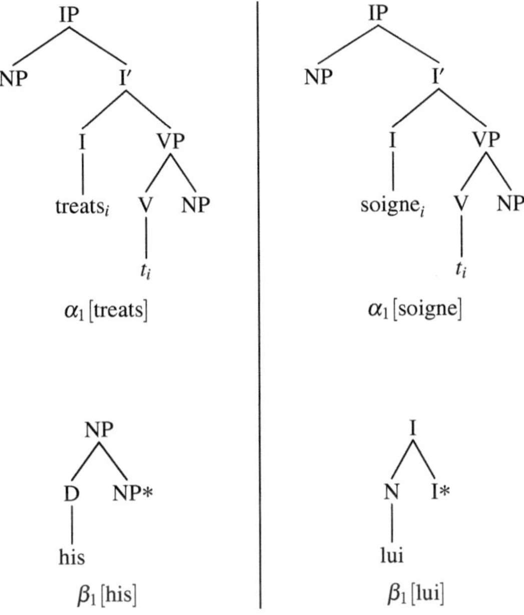

Fig. 4.6 Grammar (object-level) for English-French clitic example (sentences 4.31 and 4.32).

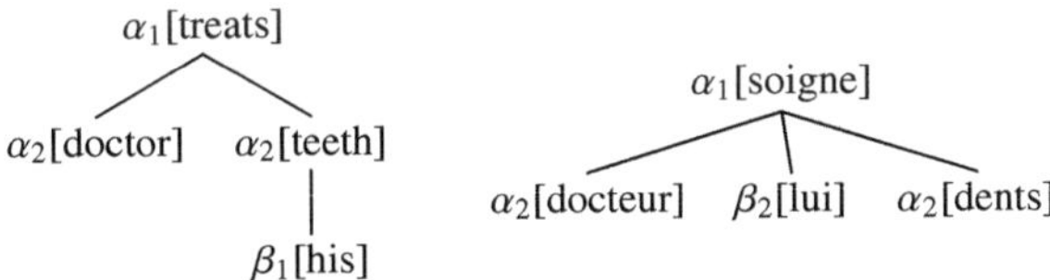

Fig. 4.7 Nonisomorphic derivations for English-French clitic example (sentences 4.31 and 4.32).

We can use a TSG to generate the nonisomorphic derivations such that the nonisomorphism is localized in a single meta-level elementary tree pair (see Figure 4.8). These trees are labeled both with names of object-level elementary trees and with nonterminal symbols. The latter are a bit of notational abuse: a node labeled with a nonterminal symbol X actually stands for the auxiliary tree with a single node $X*$. These simply serve as placeholders where meta-level adjunction can take place.

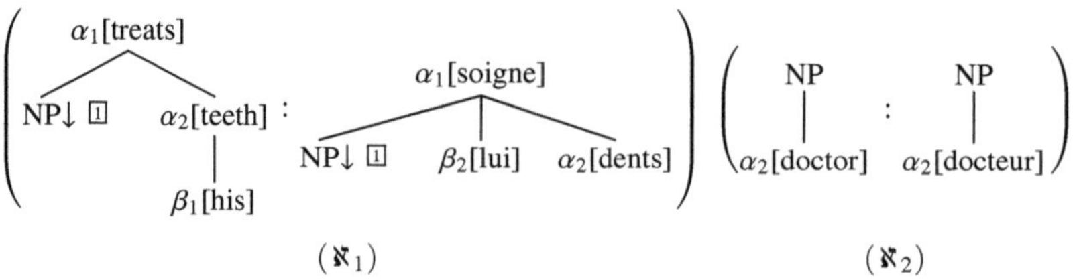

$$\left(\begin{array}{c} \alpha_1[\text{treats}] \\ \text{NP}\!\downarrow \boxed{1} \quad \alpha_2[\text{teeth}] \\ \mid \\ \beta_1[\text{his}] \end{array} : \begin{array}{c} \alpha_1[\text{soigne}] \\ \text{NP}\!\downarrow \boxed{1} \quad \beta_2[\text{lui}] \quad \alpha_2[\text{dents}] \end{array}\right) \quad \left(\begin{array}{c} \text{NP} \\ \mid \\ \alpha_2[\text{doctor}] \end{array} : \begin{array}{c} \text{NP} \\ \mid \\ \alpha_2[\text{docteur}] \end{array}\right)$$

$$(\aleph_1) \qquad\qquad (\aleph_2)$$

Fig. 4.8 Meta-level elementary tree pairs to localize nonisomorphism in English-French clitic example (sentences 4.31 and 4.32).

A more substantial extension: TAG ∘ RF-TAG

Dras and Bleam [50] consider the parallel case in Spanish, where the nonisomorphism may be unbounded because the clitic 'le' can climb up to higher verbs in the tree:

(4.33)   The doctor treats his teeth.

(4.34)   El médico le   examina los dientes.
the doctor him treats   the teeth

(4.35)   The doctor wants to (be able to...) treat his teeth.

(4.36)   El médico le   quería (poder...) examinar los dientes.
the doctor him wants able   treat   the teeth

In our own work, we have tackled a similar problem identified by Schuler [118] for translating between Portuguese and English [39]. However, the Spanish clitic-climbing case is of more interest here. One of Dras and Bleam's proposed solutions uses a meta-level grammar to map between the nonisomorphic derivations; but since TSG as a meta-level grammar cannot localize unbounded nonisomorphisms, they use RF-TAG (see Section 2.5.1), following Dras's original work on paraphrase [49].

The object-level grammar is shown in Figure 4.9. In order to generate sentences 4.35 and 4.36, we need the nonisomorphic derivations shown in Figure 4.10, assuming that 'le' must adjoin into the elementary tree of the verb to which it cliticizes. In order to generate these derivations, we need the meta-level RF-TAG shown in Figure 4.11.

4.5 Summary

Figure 4.12 summarizes the relative SGC of various synchronous grammar formalisms with respect to the domain of string and tree pairs, including some results from an earlier paper [33] not proven here. It is striking that SGC with respect to

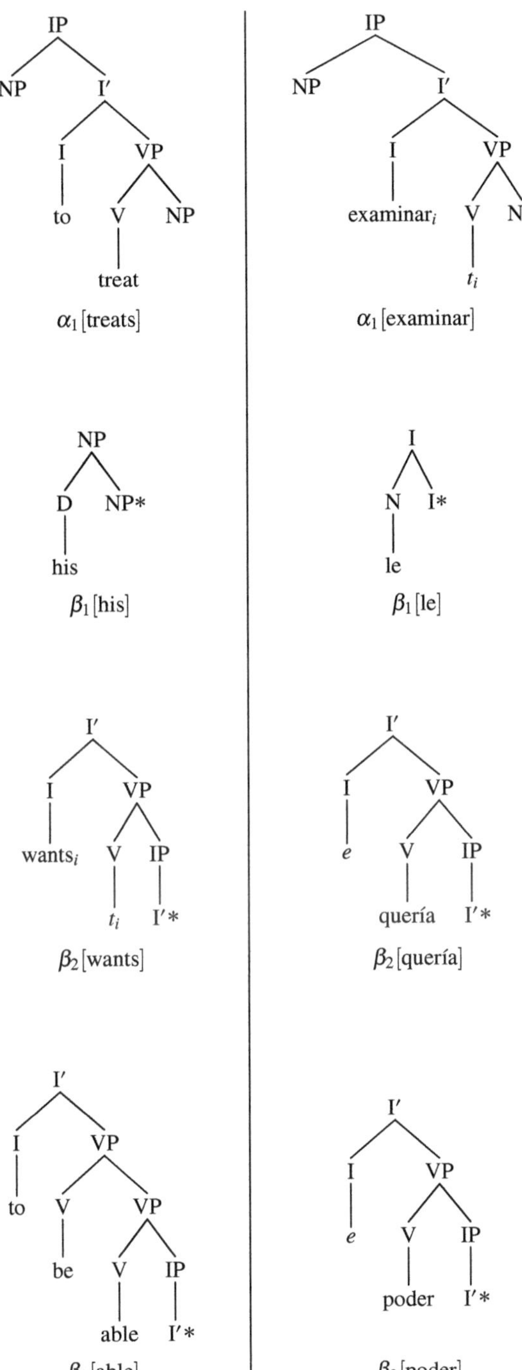

Fig. 4.9 Grammar (object-level) for English-Spanish clitic-climbing example (sentences 4.35 and 4.36).

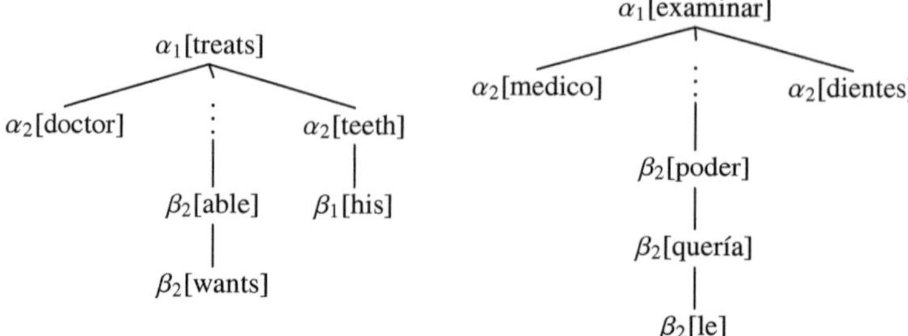

Fig. 4.10 Nonisomorphic derivations for English-Spanish clitic-climbing example (sentences 4.35 and 4.36).

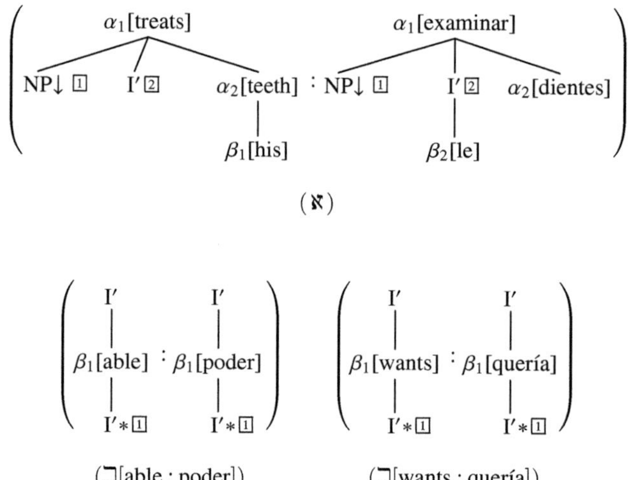

Fig. 4.11 Meta-level elementary tree pairs to localize nonisomorphism in English-Spanish clitic-climbing example (sentences 4.35 and 4.36).

the domain of string pairs classifies formalisms quite finely: inversion transduction grammar (2CFG), CFG, and RF-TAG all have different SGC with respect to the domain of string pairs even though all have the same WGC. With respect to the domain of tree pairs, the formalisms are classified even more finely.

What can we conclude? With statistical parsing, we saw that there was only a conceptual benefit gained by moving from probabilistic CFG to probabilistic TSG. But with machine translation, the advantage of formalisms beyond synchronous CFG is real. We cannot use synchronous CFG to get the same SGC as formalisms like synchronous RF-TAG.

Current research in statistical machine translation has explored the use of synchronous CFG as well as synchronous grammars which are TSGs on one side and CFGs (i.e., one-level elementary trees) on the other side. Progress on using full syn-

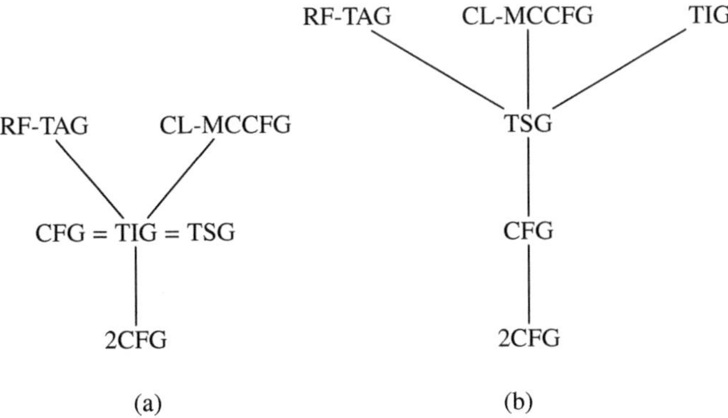

Fig. 4.12 Relative translation power of various formalisms: (a) string pairs; (b) tree pairs. An edge connecting two formalisms indicates that the higher has greater SGC than the lower. CL-MCCFG = component-local multicomponent CFG; 2CFG = rank-2 (binary branching) CFG.

chronous TSG to capture syntactic information on both sides has been slow, because nonisomorphisms between source and target syntactic structures tend to overconstrain extraction of the grammar. Several lines of research have tried to attack this problem, one of which is to extend the grammar formalism [48, 136]. It remains to be seen whether further increasing the SGC of the formalism (with respect to the domain of tree pairs) will provide all the flexibility needed to model syntactic nonisomorphisms between languages.

Chapter 5
Biological Sequence Analysis: Basics

A central problem in computational biology is analyzing genetic sequences to determine the structures of the molecules (RNAs, proteins) they code for. Since this problem is analogous to the problem in computational linguistics of describing what structural descriptions are specified by a given utterance, as first observed by Searls [120], many researchers have tried using formal grammars to analyze biological sequences as well [109, 1, 127, 106]. In this chapter we will attempt a synthesis of this family of approaches, and investigate what properties of formal grammars will determine their success.

5.1 Background

We first give an overview of some basic concepts in structural biology from a formal-language-theoretic point of view. A more thorough and conventional introduction can be found in biology textbooks [21, 7].

5.1.1 Sequences

DNA molecules are built out of nucleotides, which come in four kinds (adenine, thymine, cytosine, and guanine, or A, T, C, and G for short). Each contains a five-carbon sugar, and the only way for two nucleotides to combine is for the fifth ($5'$) carbon of one to be joined by a covalent bond to the third ($3'$) carbon of the other with a phosphate group in between (see Figure 5.1). This causes nucleotides to form into chains; the end with its $5'$ carbon free is called the $5'$ end and the end with its $3'$ carbon free is called the $3'$ end. This asymmetry is significant; the chain is always synthesized starting with the $5'$ end, for example. Thus we can think of DNA molecules as strings over $\{a, t, c, g\}$.

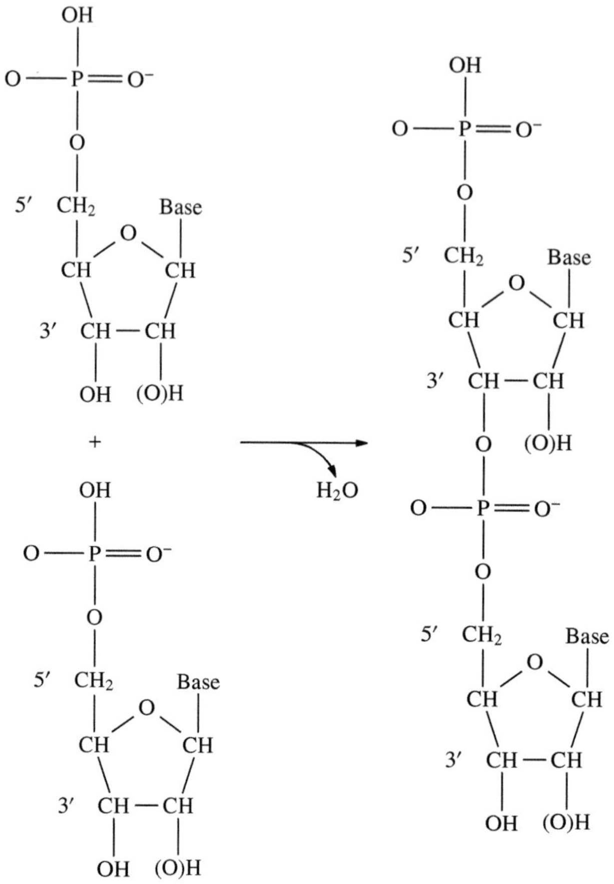

$$OH$$
$$O\!-\!P\!=\!O^-$$
$$O$$
$$5'\ CH_2 \qquad Base$$
$$O$$
$$CH \qquad CH$$
$$3'\ CH\!-\!CH$$
$$OH \quad (O)H$$

$+$

$$OH$$
$$O\!-\!P\!=\!O^-$$
$$O$$
$$5'\ CH_2 \qquad Base$$
$$O$$
$$CH \qquad CH$$
$$3'\ CH\!-\!CH$$
$$OH \quad (O)H$$

$\longrightarrow$

H_2O

$$OH$$
$$O\!-\!P\!=\!O^-$$
$$O$$
$$5'\ CH_2 \qquad Base$$
$$O$$
$$CH \qquad CH$$
$$3'\ CH\!-\!CH$$
$$O \quad (O)H$$
$$O\!-\!P\!=\!O^-$$
$$O$$
$$5'\ CH_2 \qquad Base$$
$$O$$
$$CH \qquad CH$$
$$3'\ CH\!-\!CH$$
$$OH \quad (O)H$$

Fig. 5.1 Nucleotides combining to form (a segment of) a DNA/RNA molecule. The parenthesized O is absent in DNA.

DNA is purely informational: it has no function other than to be replicated and transcribed into other alphabets. One such alphabet is that of RNA, which is similar to DNA in structure, except it uses uracil (U) instead of thymine. Thus we can think of RNA molecules as strings over $\{a, u, c, g\}$. RNA is transcribed from DNA by a base-to-base mapping ($A \to U, T \to A, C \to G, G \to C$). RNAs come in various types: transfer RNAs are about 80–100 bases long, and ribosomal RNAs are about 120–2500 bases long.

Proteins are built out of amino acids, which come in twenty kinds (see Figure 5.3). Each amino acid contains an amino group (NH_2) and a carboxyl group (COOH), and the only way for two amino acids to combine is for the amino group of one to be joined by a covalent bond to the carboxyl group of the other (see Figure 5.2). Thus in some diagrams the ends of a protein molecule are labeled N (for the nitrogen atom in the amino group) and C (for the carbon atom in the carboxyl

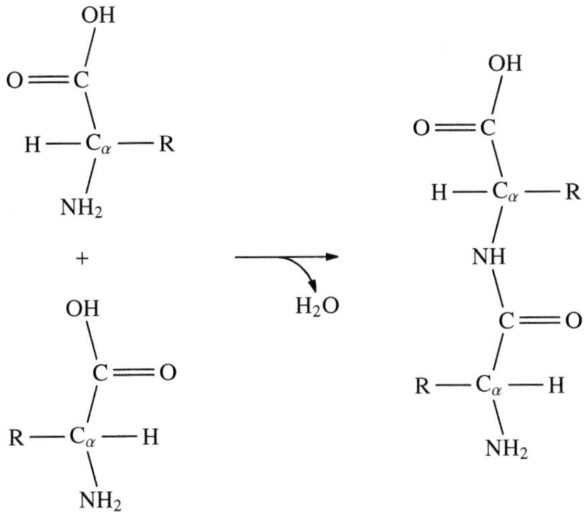

Fig. 5.2 Amino acids combining to form (a segment of) a protein. *R* indicates a side chain, which varies from amino acid to amino acid.

A Ala alanine	L Leu leucine
R Arg arginine	K Lys lysine
N Asn asparagine	M Met methionine
D Asp aspartic acid	F Phe phenylalanine
C Cys cysteine	P Pro proline
Q Gln glutamine	S Ser serine
E Glu glutamic acid	T Thr threonine
G Gly glycine	W Trp tryptophan
H His histidine	Y Tyr tyrosine
I Ile isoleucine	V Val valine

Fig. 5.3 Amino acids and their abbreviations.

group). Thus, as with DNA and RNA, we can think of proteins as strings over the set of amino acids. Proteins are transcribed from DNA via RNA by the so-called genetic code, which maps triples of nucleotides (called codons) to amino acids (see Figure 5.4). There are also start and stop codons which permit multiple proteins to be encoded in a single strand of DNA. Proteins can be quite long, exceeding 5000 amino acids in length.

5.1.2 Structures

Though RNA molecules and protein molecules have a linear structure as described above, they do not lie straight in space, because the chemical bonds in them are

	1st		2nd			3rd
		T	C	A	G	
T	Phe	Ser	Tyr	Cys		T
						C
	Leu		Stop	Stop		A
				Trp		G
C	Leu	Pro	His	Arg		T
						C
			Gln			A
						G
A	Ile	Thr	Asn	Ser		T
						C
	Met		Lys	Arg		A
						G
G	Val	Ala	Asp	Gly		T
						C
			Glu			A
						G

Fig. 5.4 The genetic code (DNA → amino acids).

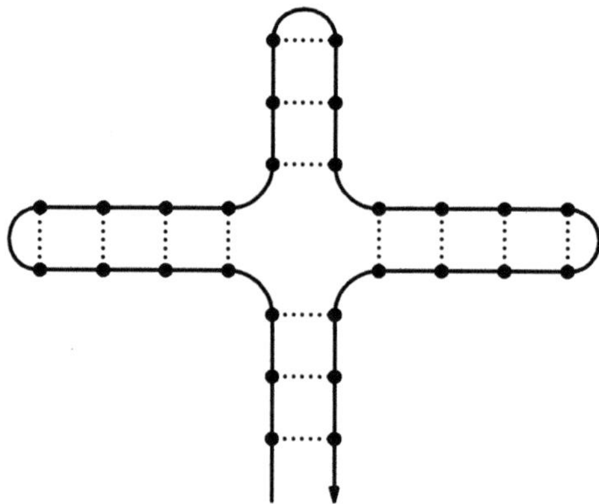

Fig. 5.5 Example RNA secondary structure.

flexible. Nor do they get bent around every which way, because *self-contacts* form between different parts of the molecule to give it a *secondary structure* (see Figure 5.5), the primary structure being the sequence itself. Distant parts of the secondary structure may come into contact to form a *tertiary structure*.

In RNA, bonds form between complementary bases: A with U, C with G.[1] Thus the secondary/tertiary structure of an RNA molecule is dependent on its primary structure. This structure gives the molecule its particular function: transfer RNAs pair codons with amino acids, and ribosomal RNAs form the machinery which assembles amino acids into proteins.

---

[1] Sometimes uracil can pair with other bases besides adenine.

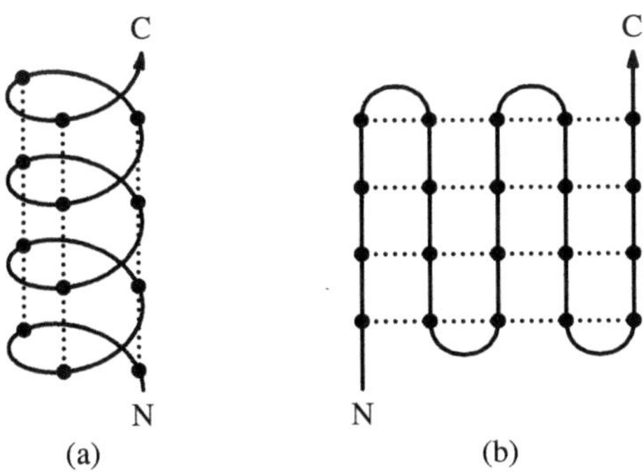

Fig. 5.6 Example protein secondary structures: (a) α-helix; (b) β-sheet.

With proteins the secondary structure is again dependent on its primary structure. Amino acids do not have complementary pairs, but do have varying properties that make some pairings more favorable than others. Protein secondary structures are observed to consist mainly of two types of substructure: *α-helices* and *β-sheets*. Other regions are known as *random coil*. In an α-helix a single region is coiled up into a helix (see Figure 5.6a); in a β-sheet several discontiguous regions are stretched flat to form a sheet (Figure 5.6b). These fold up further into a *tertiary structure*. As with RNAs, the structure of a protein determines its particular function. They perform a wide range of functions, from catalyzing biochemical reactions to giving cells their shape.

In this chapter, we assume that a molecule's structure is uniquely specified by a set of self-contacts, that is, pairs of string positions. This representation is commonly used in the structure prediction literature, where it is also known as a *polymer graph* or *contact map*.

5.2 Measuring Sequence Analysis Power

There is a long tradition of applying language-processing techniques (for example, hidden Markov models) to genetic sequences, but the use of formal grammars originated with Searls [120]. Consider the following CFG for RNA sequences:

$$
\begin{aligned}
S &\to Z \\
X &\to aZu \mid uZa \mid cZg \mid gZc \\
Y &\to aY \mid uY \mid cY \mid gY \mid \varepsilon \\
Z &\to YXZ \mid Y
\end{aligned}
\tag{5.1}
$$

A derivation of a string w, represented as a tree (see Figure 5.7), has the same shape as a secondary structure of w, because the grammar is written so that only complementary bases appear in the same rule, and CFG derivation trees have the convenient property that symbols from the same rule appear next to each other in the tree. Formal locality corresponds to spatial locality.

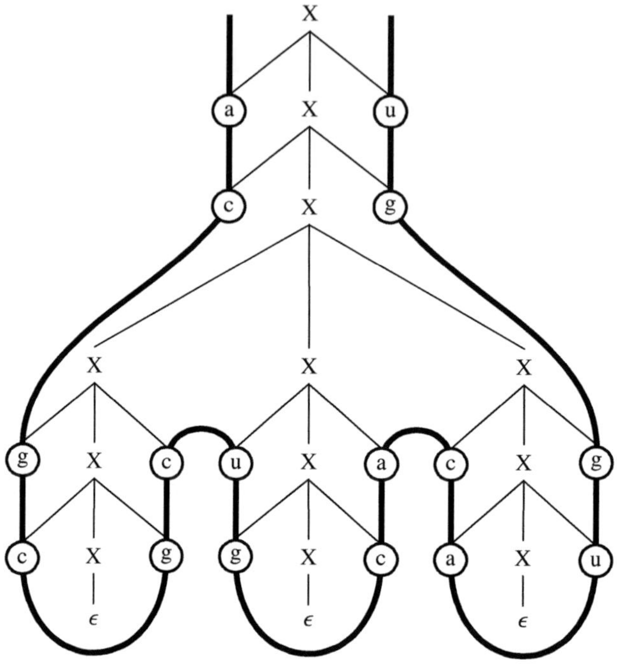

Fig. 5.7 Example CFG derivation of RNA secondary structure, with superimposed primary structure. Nonterminal symbols other than X are suppressed for clarity.

It would seem that more complex formalisms do not have this property. For example, in a TAG the adjunction operation can cause parts of an elementary tree to be stretched arbitrarily far apart.[2] But if we distinguish between spatial locality in our drawings of derivations and spatial locality in real molecules, then it becomes apparent that the former is convenient but not crucial. Even if formal locality cannot correspond to spatial locality in our drawings of derivations, they can still correspond to spatial locality in real molecules. In other words, derivations can still describe molecules even if their drawings don't look very much like them.

This gives us the following locality constraint: two nonadjacent monomers can contact only if their corresponding symbols were generated in the same derivation step. All uses of formal grammars to model biological molecules that we know of are based on this principle, though with variations and sometimes only implicitly.

---

[2] However, Rogers [108] explores the use of three-dimensional trees to represent derivations of tree-adjoining grammars, and higher-dimensional trees for still more complex formalisms. In a tree-adjoining grammar defined on three-dimensional trees, there is no stretching.

In Searls' original treatment [120] and that of Uemura et al. [127], any two bases appearing in the same right-hand side are assumed to be in contact. Rivas and Eddy [106] use diagrams reminiscent of Joshi's links [67]. It is a deficiency of the model of Abe and Mamitsuka [1] that they do not specify self-contacts on their elementary structures, with the result that a single derivation can correspond to multiple structures.

Chen and Dill [29] do not use a grammar at all, but their model can be recast as a CFG [37]. Strictly speaking, this grammar does not generate strings of monomers but strings of covalent bonds. Below is a simplified version of their grammar:

$$S \rightarrow Z \mid X$$
$$X \rightarrow \bullet Z \bullet$$
$$Y \rightarrow -Y \mid -$$
$$Z \rightarrow YX \mid XZ \mid YXZ \mid Y$$

$$(5.2)$$

where $-$ is a terminal symbol (representing a covalent bond) and $\bullet$ is not a symbol, but simply a potential endpoint for a link. The advantage of using an alphabet of covalent bonds here is that multiple rules can create links on one monomer. Indeed, this grammar can generate an arbitrary number of links on a single monomer. In our experiments in Chapter 6, we use a grammar of this type; however, since grammars like (5.1) are more intuitive, we will use it as the basis for our examples below.

The grammars used for modeling molecules typically generate the language Σ^*, since we are generally not interested in accepting or rejecting molecules as "grammatical," but determining their structures. Weak generative capacity, then, is not useful as an indicator of the usefulness of a grammar formalism for modeling molecules. What matters is a formalism's derivational generative capacity (1.3), which, as we have already seen, is a measure of ability to generate linked strings.

5.3 Linked Grammars for Sequence Analysis

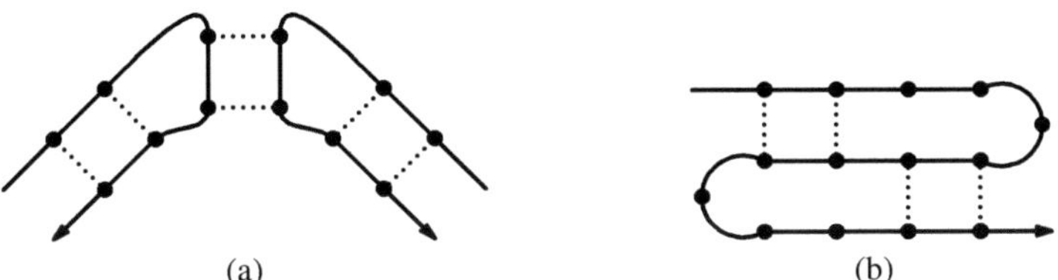

Fig. 5.8 Example RNA tertiary structures: (a) kissing hairpins; (b) pseudoknot.

In the previous section we showed how CFG has been used to model RNA and protein structures. Does CFG have enough DGC? It is commonly said that CFG cannot generate crossing dependencies. This is not strictly true, since even a single production can have crossing dependencies:

$$X \rightarrow a\,b\,c\,d \tag{5.3}$$

Nevertheless, there are patterns of crossing self-contacts which occur in nature which are provably not generable by CFG (or by Chen and Dill's model). For example, helices involve unbounded series of short crossing self-contacts. RNA tertiary structures involve long-distance crossing self-contacts, for example, *kissing hairpins* and *pseudoknots* (Figure 5.8). Finally, protein β-sheets involve patterns of crossing self-contacts that are well beyond the power of CFG.

5.3.1 Squeezing DGC

We first consider the use of grammar formalisms coverable by CFG which have greater DGC than CFG.

Alpha-helices

Helices have crossing self-contacts of a very limited type. Below we will be interested in helices in which the $(2i)$th monomer is in contact with the $(2i-3)$rd monomer, as this is how helices appear on a square lattice (see Figure 5.9a). As an exercise, we may generate such helices with the following multicomponent CFG (in the sLMG notation of Section 2.5.2):

$$
\begin{aligned}
S(a_1 a_2 y a_3 x) &:- H(y:x) \quad a_i \in \Sigma \\
H(a_1 : y a_2 x) &:- H(y:x) \quad a_i \in \Sigma \\
H(a, \varepsilon) & \qquad\qquad\ \ a \in \Sigma
\end{aligned}
\tag{5.4}
$$

Since this grammar is component-local, it can be covered by a CFG (indeed, by a right-linear CFG or finite-state automaton). Its derivations can be faithfully represented as derivations of the dissolved grammar (for example, see Figure 5.9b).

Proposition 7. *The linked language L generated by the above component-local multicomponent CFG cannot be generated by any CFG.*

Proof. Suppose L is generated by a CFG G. If a linked string generated by a CFG contains two crossing dependencies as in (5.3), all four terminal symbols involved must come from the same production. By induction, this means that all the terminals in every string of L must come from the same production. Since L is infinite, G must be infinite, which is a contradiction.

In a previous paper [33] we showed that component-local multicomponent CFG can even generate linked languages that TAG cannot.

Other kinds of helices can be modeled as well; in particular, if we use an alphabet of covalent bonds (instead of monomers), we can model helices in which the ith monomer is in contact with the $(i-4)$th monomer, as in real α-helices. All of this is somewhat academic, however, because it is easier just to write down the cover grammar directly. For example, the Zimm-Bragg model [137] is a standard theory of the helix-coil transition and can be thought of as a Markov chain.

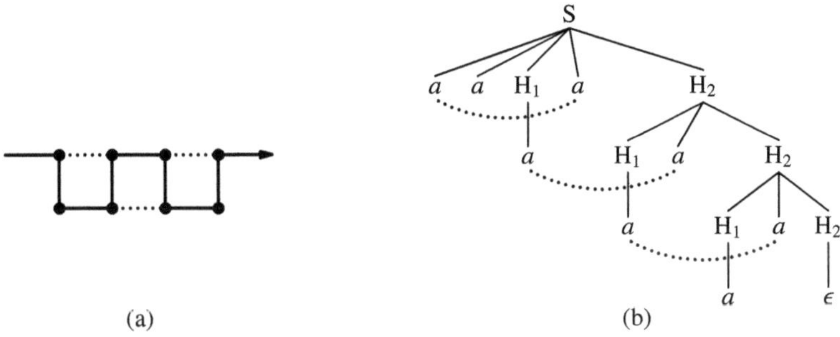

(a)   (b)

Fig. 5.9 (a) α-helix as modeled by square lattice; (b) α-helix represented by derivation of grammar (5.4).

Limited RNA tertiary interactions

The following results show how regular form TAG (see Section 2.5.1) can be used to model limited tertiary interactions in RNAs.

Lemma 2 (linked pumping lemma). *Let L be a linked string set generated by a CFG (or component-local multicomponent CFG). Then there is a constant n such that if $\langle z; \sim_z \rangle$ is in L and $|z| \geq n$, then z may be rewritten as $uvwxy$, with $|vx| > 0$ and $|vwx| \leq n$, such that for all $i \geq 1$, there is a relation $\sim_z^i$ such that $\langle uv^i wx^i y; \sim_z^i \rangle$ is in L and $\sim_z^i$ does not relate any positions in w to any positions in u or y.*

Proof. The proof is analogous to that of the standard pumping lemma [59, pp. 125–127]. However, a CFG cannot be put into Chomsky normal form without changing its DGC, so we let $n = m^k$ instead of 2^k, where k is the size of the nonterminal alphabet and m is the longest right-hand side or the number 2, whichever is greater.

The key difference from the standard proof is the observation that since, for each i, the derivation of $uv^i wx^i y$ can be written as

$$S \overset{*}{\Rightarrow} uAy \overset{*}{\Rightarrow} uv^i Ax^i y \overset{*}{\Rightarrow} uv^i wx^i y$$

for some nonterminal A, no position in w can be contributed by the same derivation step as any position in u or y.

The generalization to the multicomponent case is straightforward, since a component-local multicomponent CFG G can be converted into a CFG G' which generates the same trees. G' will not generate the same linked strings as G; nevertheless, the equivalence relations generated by G can only relate terminal instances which are first cousins in the derived tree, so for $i \geq 1$, it remains the case that no position in w is related to any position in u or y.

Proposition 8. *The following linked string set is generable by an RF-TAG but not by any CFG, nor indeed by any component-local multicomponent CFG:*

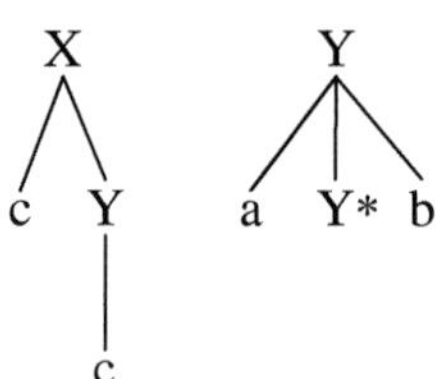

$$L = \left\{ \text{c a a} \cdots \text{a c b} \cdots \text{b b} \right\}$$

Proof. The following RF-TAG generates L:

But suppose L is generated by some CFG G. For any n given by the linked pumping lemma, let $z = \text{ca}^n\text{cb}^n\text{c}$ satisfy the conditions of the pumping lemma. It must be the case that v and x contain only a's and b's, respectively, or else $uv^i wx^i y \notin L_1$. But then u, w, and y would each have to contain one of the c's, and since the c's are all related, this contradicts the pumping lemma.

Grammars of this type could be used for structures in which all but a bounded number of self-contacts are nested. For example, in a cloverleaf structure (Figure 5.5), the hairpins may "kiss" (Figure 5.8a), forming a small number of self-contacts crossing over an unbounded number of nested self-contacts. If the number of such self-contacts is indeed bounded, we can write an RF-TAG similar to the one above to generate them (see Figure 5.10).

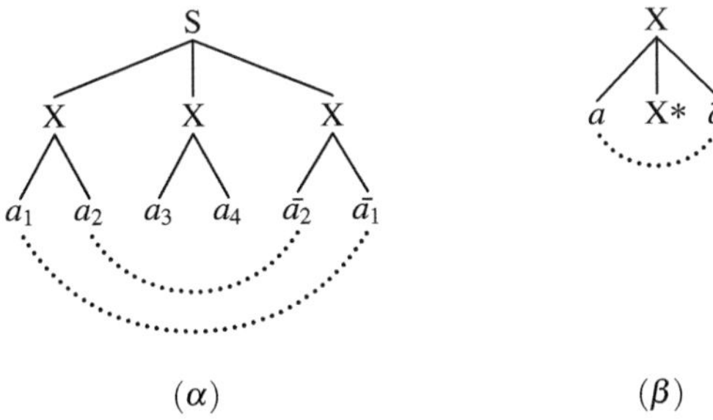

Fig. 5.10 RF-TAG for cloverleaf with kissing hairpins. The initial tree α generates the loops (here fixed to two monomers each), and β generates the stem regions.

5.3.2 Beyond CFG

We now examine some existing attempts to apply formalisms beyond CFG to more complex structures.

Pseudoknots

Uemura et al. [127] use a grammar similar to the one shown in Figure 5.11 to generate pseudoknot structures. A pseudoknot is generated by repeatedly adjoining β_1 into α, then repeatedly adjoining β_2 into the result. Their grammar belongs to an $\mathcal{O}(n^5)$-time parseable subclass of TAG, which has been conjectured [71] to be equivalent to the TAG restriction of Satta and Schuler [110].

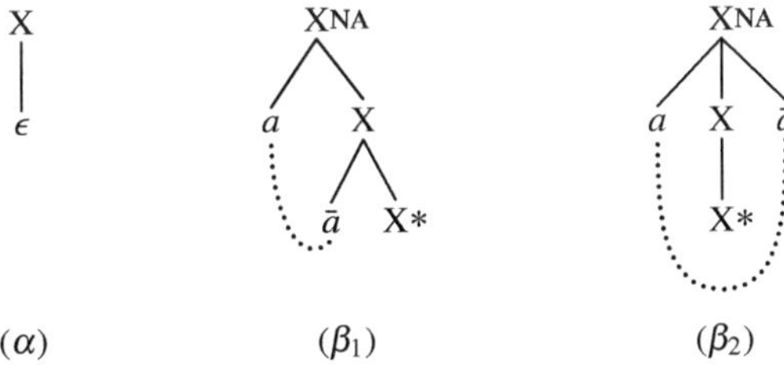

Fig. 5.11 TAG fragment for pseudoknots, adapted from the grammar of Uemura et al. [127].

Rivas and Eddy [106] define a formalism called crossed-interaction grammar. Their definition of this formalism is unclear in many places and seems to be as powerful as type-0 grammars; but a little exegesis shows that what they had in mind is something equivalent to linear sLMG. They then use a grammar similar to the one shown in Figure 5.12 to generate pseudoknot structures. The first three rules generate basic pseudoknots just as the TAG of Figure 5.11. Like a TAG, Rivas and Eddy's grammar has a maximum arity of two and a maximum branching factor of two and is therefore parsable in $\mathscr{O}(n^6)$ time, but it lies outside the power of TAG. For example, the last rule allows arbitrary-length chains of hairpins to be generated (Figure 5.13 shows a chain of four), whereas TAG can only build such chains of up to three hairpins.

$$W(x_1y_1x_2y_2) :- W_H(x_1,x_2), W_H(y_1,y_2)$$
$$W_H(ax : y\bar{a}) :- W_H(x,y) \qquad\qquad a \in \Sigma$$
$$W_H(\varepsilon,\varepsilon)$$
$$W_H(x_1y_1x_2,y_2) :- W_H(x_1,x2), W_H(y_1,h_2)$$

Fig. 5.12 sLMG fragment for pseudoknots, adapted from the grammar used by Rivas and Eddy [106], as a linear sLMG.

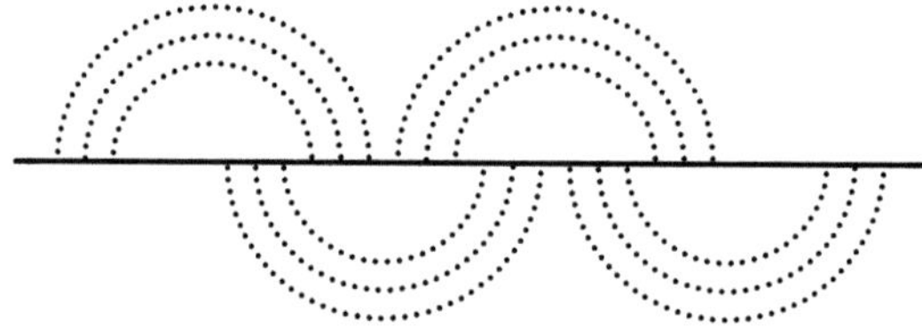

Fig. 5.13 Chain of four hairpins beyond the power of TAG.

Beta-sheets

Abe and Mamitsuka [1] use a formalism called ranked node-rewriting grammar (RNRG) to generate β-sheets. RNRG is essentially TAG with multiple foot nodes on elementary trees; for the present discussion, it is enough to note that RNRG with a branching factor of one is equivalent to linear sLMG with a maximum branching factor of one. Figure 5.14 shows a grammar for generating a β-sheet of eight alternating strands. Set-local multicomponent TAG offers a similar solution; Figure 5.15 shows a grammar to generate the five-strand sheet of Figure 5.6b.

The difficulty is that parsing of these grammars is exponential in the number of strands per sheet. Moreover, every grammar imposes some upper bound, so that there is no single grammar that can generate all β-sheets. For this reason, approaches of this type appear to be prohibitively expensive.

$$S(a_1x_1b_1, a_2x_2b_2, a_3x_3b_3, a_4x_4b_4) :- S(x_1, x_2, x_3, x_4) \qquad a_i, b_i \in \Sigma$$
$$S(\varepsilon, \varepsilon, \varepsilon, \varepsilon)$$

Fig. 5.14 sLMG fragment for β-sheet, adapted from one of the grammars of Abe and Mamitsuka [1].

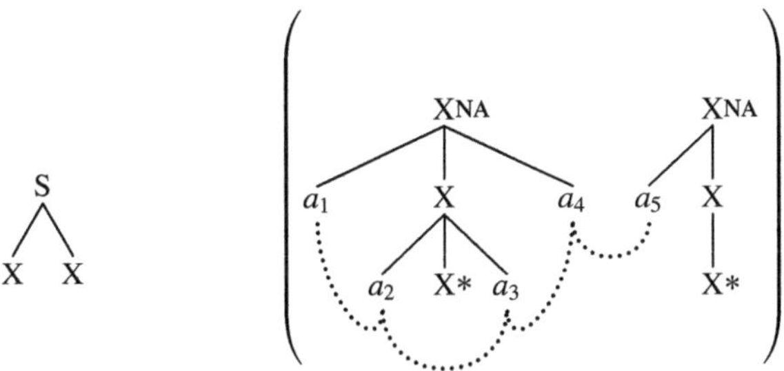

Fig. 5.15 Set-local multicomponent TAG for protein β-sheet of Figure 5.6b.

A second problem is that this analysis is susceptible to a kind of spurious ambiguity in which a single structure can be derived in multiple ways. For example, consider Figure 5.16. In order to generate the β-sheet (a), we need trees like (b) and (c). But either of these trees can be used by itself to generate the β-sheet (d). The grammar must make room for the maximum number of strands, but when it does not use all of it, ambiguity can arise. It should be possible to carefully write the grammar to avoid much of this ambiguity, but we have not been able to eliminate all of it even for the single-component TAG case.

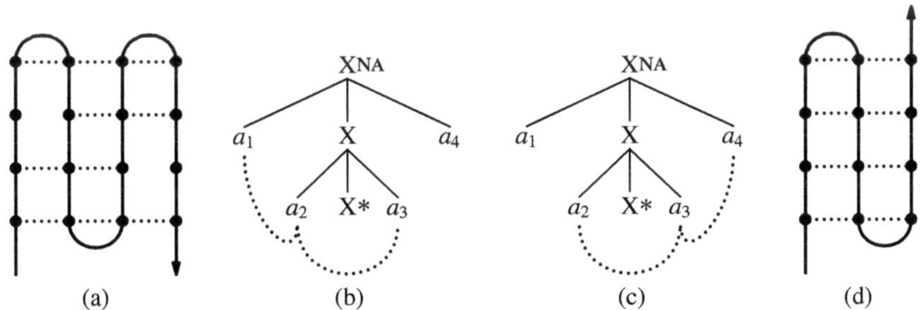

Fig. 5.16 Illustration of spurious ambiguity in a multicomponent TAG.

5.4 Computing Probabilities and Partition Functions

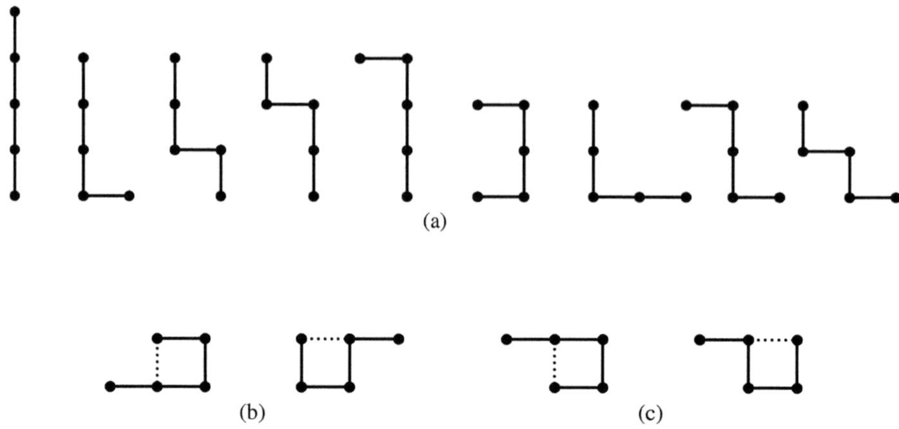

Fig. 5.17 All possible conformations of a 5-mer on a square lattice (modulo rotational and reflectional symmetry), grouped according to structures, (a), (b), and (c).

Just as with natural language parsing, a grammar can assign many derivations to a single string, and we would like some measure of the relative importance of each of them—for example, a probability distribution. Attempts have been made to estimate probabilities from databases [109, 1]. However, maximizing the likelihood of a database would not seem to make much sense, since the database does not represent a uniform sample of any naturally occurring distribution (as a text corpus arguably does). Maximizing the conditional likelihood might be a sounder approach (Hockenmaier, p.c.). But physical theory provides a more principled alternative. If we have some way of calculating the energies of structures, minimizing the energy will give us the most stable structure. In fact, statistical physics can tell us the full probability distribution over structures in terms of their energy.

Consider a set of many identical molecules, each in a particular *conformation*, or arrangement in physical space. In the HP lattice model [80], conformations are represented by self-avoiding walks on a lattice (see Figure 5.17). Note that there can be more than one conformation corresponding to a single structure; for example, in Figure 5.17, all the conformations in the first row have the same structure. The maximum-entropy probability distribution over the conformations of these molecules, subject to the constraint that their total energy must be constant, is known as the *Boltzmann distribution*:

$$P_j = \frac{e^{-E_j/kT}}{Q} \tag{5.5}$$

where j ranges over conformations, E_j is the energy of conformation j, T is the temperature, k is Boltzmann's constant, and

$$Q = \sum_j e^{-E_j/kT} \tag{5.6}$$

is known as the *partition function*. Since we are more interested in distributions over structures than conformations, we may regroup both the Boltzmann distribution and the partition function according to structure:

$$P_j = \frac{\Omega_j e^{-E_j/kT}}{Q} \tag{5.7}$$

$$Q = \sum_j \Omega_j e^{-E_j/kT} \tag{5.8}$$

where j now ranges over structures, and Ω_j is the number of conformations with structure j.

When T is low, the Boltzmann distribution says that the most probable structure will simply be the one(s) with the lowest energy: let $E_{min} = \min E_j$, then

$$\lim_{T \to 0} P(j) = \lim_{T \to 0} \frac{\Omega_j e^{-E_j/kT}}{\sum_{j'} \Omega_{j'} e^{-E_j/kT}} \tag{5.9}$$

$$= \lim_{T \to 0} \frac{\Omega_j e^{-E_j/kT + E_{min}/kT}}{\sum_{j'} \Omega_{j'} e^{-E_j/kT + E_{min}/kT}} \tag{5.10}$$

$$= \begin{cases} \dfrac{\Omega_j}{\sum_{j' \text{ s.t. } E_{j'} = E_{min}} \Omega_{j'}} & \text{if } E_j = E_{min}, \\ 0 & \text{otherwise.} \end{cases} \tag{5.11}$$

When T is high, the Boltzmann distribution predicts a uniform distribution of conformations, therefore giving preference to structures with more conformations.

The partition function gives the *effective accessibility* of each conformation (0 being fully inaccessible, 1 being fully accessible) and turns out to be the more useful object to compute. All the statistical-mechanical properties of a system, including the Boltzmann distribution, can be computed from it. We can use it, for example, to understand the folding process of a molecule, or the changes it undergoes under varying conditions, which can shed further light on its function.

How do we compute the partition function using a weighted grammar? If we can assign quantities ω_π and ΔE_π to each elementary structure of the grammar π such that for every structure j built out of elementary structures $\pi_1, \ldots, \pi_n$,

$$\prod_i \omega_{\pi_i} \approx \Omega_j \tag{5.12}$$

$$\sum_i \Delta E_{\pi_i} \approx E_j \tag{5.13}$$

and assign the weight $\omega e^{\Delta E/kT}$ to each elementary structure, then the weight of the derivation of j will be

$$\prod_i \omega_{\pi_i} e^{-\Delta E_{\pi_i}/kT} \approx \Omega_j e^{-E_j/kT} \tag{5.14}$$

The problem, then, is to design the grammar such that the energy increments (ΔE_π) and conformation counts (ω_π) can be estimated accurately.

Calculating energies

Attaching energies to rules is common practice among previous grammatical approaches to structure prediction. Here we describe a particularly simple energy model. In the HP model [80], monomers are classified as either hydrophobic (h) or polar (p), and hh contacts are favorable. That is, the energy of an hh contact is $\varepsilon < 0$, and the energy of other self-contacts is zero. Therefore, we can adapt grammar (5.1) as follows, letting $q_{hh} = e^{-\varepsilon/kT}$:

$$\begin{aligned}
S &\xrightarrow{1} Z \\
X &\xrightarrow{q_{hh}} hZh \\
X &\xrightarrow{1} hZp \mid pZh \mid pZp \\
Y &\xrightarrow{1} hY \mid pY \mid \varepsilon \\
Z &\xrightarrow{1} YXZ \mid Y
\end{aligned} \tag{5.15}$$

(We use the factor $q_{hh} = e^{-\varepsilon/kT}$ here instead of the energy ε itself, so that the weights can be multiplied instead of added, for consistency with the other grammars in this section.) A parser that computes the minimum-weight derivation of a string under this grammar would compute the native structure of the corresponding molecule.

Counting conformations

Attaching conformation counts to grammar rules has not been explored previously to our knowledge. Chen and Dill [28] use a matrix computation to estimate conformation counts in polynomial time by dividing each structure into substructures and ignoring excluded volume between substructures. That is, the substructures are counted separately, and then the counts are multiplied together. The grammar can check for collisions between substructures to the extent that the shape of a substructure can be finitely encoded; but in general collisions between substructures are ignored. Their algorithm may alternatively be viewed as a parser for an implied grammar [37]. We can similarly add conformation counts to grammar (5.15):

$$S \xrightarrow{C(l-1)} Z^l$$

$$X \xrightarrow{\frac{1}{4}U(l)q_{hh}} hZ^lh$$

$$X \xrightarrow{\frac{1}{4}U(l)} hZ^lp \mid pZ^lh \mid pZ^lp$$

$$Y^{l+1} \xrightarrow{1} hY^l \mid pY^l$$

$$Y^0 \xrightarrow{1} \varepsilon$$

$$Z^{k+l+1} \xrightarrow{1} Y^kXZ^l$$

$$Z^l \xrightarrow{1} Y^l$$

$$(5.16)$$

where the superscripts are counters used for measuring lengths. Y generates open chains and X generates closed loops, and Z generates combinations of the two, counting the latter as having length one. When an X forms a closed loop out of a Z of length l, it multiplies in a conformation count of $\frac{1}{4}U(l)$, where $U(l)$ is the number of neighbor-avoiding loops of length l on the two-dimensional lattice (l even). When S forms the whole molecule out of a Z of length l, it multiplies in a conformation count of $C(l-1)$, where $C(l)$ is the number of neighbor-avoiding walks of length l. For large l we use approximations of the form $A\mu^l l^{\gamma-1}$ [85]. For $U(l)$, we use $A = 1.3$, $\gamma = \frac{11}{32}$; for $C(l)$ we use $A = 0.034$, $\gamma = 0.5$; for both formulas we use $\mu = 2.3$. These values were chosen to approximate the results of exact enumeration; more precise or more theoretically-motivated values would be desirable.

Computing partial sums of the partition function

A weighted derivation forest lets us compute the total weight Q of the forest (using the Inside algorithm), or the weight of a single derivation (corresponding to a single structure). However, we may want to further group structures into bins in various ways and compute the total weight of each bin. For example, we might want to group structures according to their energy level and ask what the total contribution of each energy level is. Or we might group structures according to how many self-contacts are present or not present in the native structure [30].

To do this, we incorporate the bins into the nonterminal alphabet. For example, to group the derivations of grammar (5.1) by the number of self-contacts, we would write:

$$S^n \rightarrow Z^n$$

$$X^{n+1} \rightarrow hZ^n h \mid hZ^n p \mid pZ^n h \mid pZ^n p$$

$$Y^0 \rightarrow hY^0 \mid pY^0 \mid \varepsilon \tag{5.17}$$

$$Z^{m+n} \rightarrow Y^0 X^m Z^n$$

$$Z^0 \rightarrow Y^0$$

This grammar has multiple start symbols S^n, one for each bin. The total weight of S^n is the total weight of states with n self-contacts. Using this grammar we can also calculate the mean and variance of the number of self-contacts, or the most likely structure for each number of self-contacts. The rule weights come from grammar (5.16); combining these two grammars, we get:

$$S^n \xrightarrow{C(l-1)} Z^{l,n}$$

$$Y^{l+1,0} \xrightarrow{1} hY^{l,0} \mid pY^{l,0}$$

$$Y^{0,0} \xrightarrow{1} \varepsilon$$

$$Z^{k+l+1,m+n} \xrightarrow{1} Y^{k,0} X^m Z^{l,n} \tag{5.18}$$

$$Z^{l,0} \xrightarrow{1} Y^{l,0}$$

$$X^{n+1} \xrightarrow{\frac{1}{4}U(l)q_{hh}} hZ^{l,n}h$$

$$X^{n+1} \xrightarrow{\frac{1}{4}U(l)} hZ^{l,n}p \mid pZ^{l,n}h \mid pZ^{l,n}p$$

Parsing CFG typically has time complexity $\mathcal{O}(|G|n^3)$. The fourth rule schema above is the one which has the most instantiations: $\mathcal{O}(n^2 B^2)$, where B is the number of bins, for two length indices and two bin indices. However, l in this rule does not contribute to parsing complexity because it is always equal to the width of the span of the Y; therefore this grammar can be parsed in time $\mathcal{O}(n^4 B^2)$. Note that B is the number of bins *for a given string*, which needs to be finite. If the bins are energy levels, they are all linear combinations (with integer coefficients) of the ΔE's, which are drawn from a fixed, finite set. If there is a number x such that each ΔE can be expressed as an integer multiple of x (it suffices for the ΔE's to be all rational), then B will be linear in the number of self-contacts. If the number of self-contacts a single terminal can participate in is bounded (in this case, it is bounded to one), then $B \in \mathcal{O}(n)$, giving an overall complexity of $\mathcal{O}(n^6)$. Similar reasoning fixes asymptotic upper bounds on other binning schemes as well.

Implementation details

In practice it is more efficient to calculate the partition function (including energies, lengths, counts, and bins) offline, after discarding chart items which are not part of a complete derivation. Since the nonterminal indices for lengths and bins do not affect grammaticality, we can first parse with grammar (5.1) and then reparse only the resulting forest with grammar (5.18).

5.5 Summary

In this chapter we have provided a synthesis of current research in the application of formal grammars to biological sequence analysis. We have characterized the ability of grammar formalisms to model secondary/tertiary structures by their DGC, and introduced a few novel ways of using extra DGC to model more complex structures. Finally, we have shown how to use extended weights in a grammar to compute partition functions, thus reformulating Chen and Dill's nongrammatical model as a weighted CFG. In the next chapter we explore the use of the technique of *intersection* to extend this model to more complex structures like bundles of α-helices, or possibly β-sheets.

Chapter 6
Biological Sequence Analysis: Intersection

Another strategy for obtaining more SGC out of a grammar formalism is to combine multiple grammars into a single system which accepts the intersection of the languages accepted by the component grammars, and which assigns to each string the *unification* (in some sense) of the structural descriptions assigned by the component grammars. This technique has not received much attention in computational linguistics, probably because linguistic structures tend to be hierarchical, and it is not very clear how to unify multiple hierarchical structures into a single one. With molecular structures, on the other hand, there is a straightforward way of unifying two linkings of a string: simply form the union of the links. In this chapter we discuss the strengths and weaknesses of several variants of this strategy.

6.1 Intersecting CFLs and CFLs: a Critique

The simplest way to specify the intersection of two languages $L = L_1 \cap L_2$ is to provide separate grammars for L_1 and L_2. A string is recognized as belonging to L just in case it is recognized as belonging to both L_1 and L_2. There is no interaction between the two grammars at the level of their derivations or structural descriptions; the only interaction is that each filters the other's generated strings.

Context-free languages are not closed under intersection [59, pp. 134–135]. This suggests the possibility of using two or more CFGs to recognize a language beyond the power of CFG. Brown and Wilson [23] propose just this approach for RNA pseudoknots. They observe that $\{a^m g^* u^m c^*\}$ and $\{a^* g^n u^* c^n\}$ are context-free languages, but their intersection is the non-context-free language $\{a^m g^n u^m c^n\}$. This language is reminiscent of a set of pseudoknots: the m a's and u's form one hairpin, and the n g's and c's are the other. Therefore this would seem to be an efficient way of modeling pseudoknots.

However, in order for the pseudoknot to be well-formed, the two hairpins must interlock without colliding. That is, the base pairings must cross, but no two pairings should involve the same base. But the only reason the above example achieves this is because one hairpin has only a's and u's and the other has only c's and g's—that is, each symbol indicates overtly which hairpin it belongs to. For real molecules, both component grammars would have to generate at least all possible hairpins, or $\{vww^R x\}$. In that case there would be no way of preventing the component grammars from missing each other or colliding.

Brown and Wilson recognize that there is a problem, but it is not clear whether they appreciate how serious it is. Their solution is to employ a special parsing strategy that uses the results of parsing with the first grammar to constrain the parse with the second; then the string is reparsed with the first, then again with the second. This procedure works only for their pair of grammars and only approximates the desired computation.

The root of the problem is that intersection only operates on strings, not structural descriptions. It allows parallel structural descriptions to be derived independently, then filters them on the basis of their string yields. The above example attempts to harness this filtering to generate only well-formed pseudoknots, but in order to do so it assumes that there is more information in the string languages than there really is.

6.2 Intersecting CFGs and Finite-State Automata

Suppose, however, that G_1 is a CFG and G_2 is a right-linear CFG, or a grammar that can be covered by a right-linear CFG. Since G_2 generates a regular language and CFLs are closed under intersection with regular languages, it is possible to construct a new CFG $G_\cap$ that generates $L(G_1) \cap L(G_2)$. Though there is no increase in WGC as in the previous section, there is still an increase in SGC, because the resulting system assigns to each string the superposition of its structural descriptions assigned by G_1 and G_2.[1] The advantage of this approach is that the two grammars are much easier to control when encapsulated into a single grammar than in the previous section.

In Section 5.3.1 we mentioned the Zimm-Bragg model of α-helices [137], which gives partition functions for conformations with local self-contacts. Chen and Dill's model gives partition functions for conformations with nested nonlocal self-contacts, but thus far it has been impossible to compute partition functions for chain molecules having both local *and* nonlocal interactions, as in bundles of helices (Figure 6.1).

But the Zimm-Bragg model is formally a weighted finite-state automaton: every helix unit preceded by a coil unit has weight σs, and every helix unit preceded by a helix unit has weight s (Figure 6.2). Moreover, this automaton could be thought of as a cover grammar for a component-local multicomponent CFG for helices (Section 5.3.1). And we have already shown that Chen and Dill's model is equivalent to a weighted CFG. Therefore, we can use the machinery of formal grammars to com-

---

[1] Of course, $G_\cap$ itself has no more power than an ordinary CFG; it is the combination of G_1 and G_2 that has greater SGC. We leave for future work the question of when this combination can be covered by a CFG.

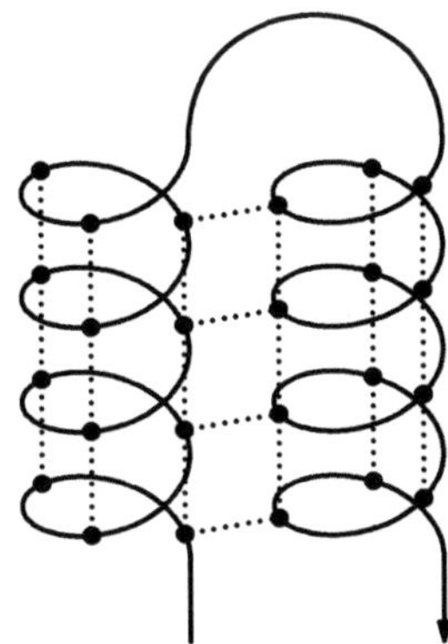

Fig. 6.1 Two-helix bundle.

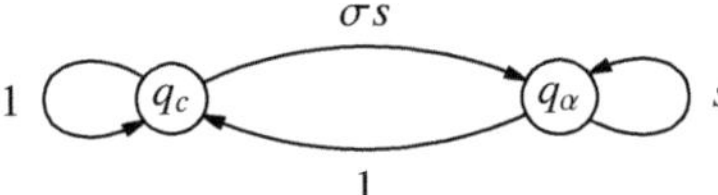

Fig. 6.2 The Zimm-Bragg model as a weighted finite-state automaton.

bine these two models easily. The combined system would generate linked strings representing two-helix bundles, and the constructed weighted CFG would correctly calculate the partition function.

6.2.1 Integrating the Zimm-Bragg Model and the HP Model

Before integrating the Zimm-Bragg model into Chen and Dill's model, we must first adapt it to the HP lattice model which underlies Chen and Dill's model. This will apply both to our final grammar-based model as well as the exact enumeration we will evaluate it against.

In a real α-helix, for each i, the ith monomer is in contact with the $(i-4)$th monomer, and between the ith and $(i-1)$st monomer, there are two bond angles (like hinges) which must be frozen into the correct shape. The Zimm-Bragg model models the self-contacts by giving each an energy of ε_s, and it models the freezing by giving a conformation count of $r < 1$ to each bond angle, representing the relative lack of conformational freedom of a helix relative to random coil. Therefore the first turn of the helix should get a weight of $r^6 e^{-\varepsilon_s/kT}$. The r^6 factor is for the six frozen bond angles between the first through fourth monomers, and the $e^{-\varepsilon_s/kT}$ for the self-contact between the first and fourth monomers. Then each subsequent monomer, because it freezes two more bond angles and adds one more self-contact, gets a weight of $r^2 e^{-\varepsilon_s/kT}$. The simplest version of the model collapses the whole first turn into the first monomer; thus the first monomer gets a weight of σs, where $\sigma \approx r^4$ and $s \approx r^2 e^{-\varepsilon_s/kT}$, and subsequent monomers get a weight of s.

Now on a square lattice, a helix is modeled as shown in Figure 5.9a. This is not completely accurate (cf. Figure 5.6a): only every other monomer creates a new self-contact, and the $(2i)$th monomer is in contact with the $(2i - 3)$rd monomer. Nevertheless, we still give every monomer an energy of ε_s, plus an energy of ε_{hh} for every hh contact.

Since we explicitly count conformations on a lattice, we are already counting random coil as having more conformations than helix—by a factor of approximately μ (the connective constant from Section 5.4) per monomer. Ideally, then, the factor r would be superfluous. In that case we would simply give each helix unit a weight of $s = e^{-\varepsilon_s/kT}$. However, because the lattice model does not match reality very exactly, we keep a correction factor on s: $s = s_0 e^{-\varepsilon_s/kT}$, where $s_0 \approx \mu r^2$. Moreover, in the lattice model the first turn is no more difficult to form than subsequent turns. Therefore we must retain the σ parameter as well.

To summarize, the factors contributing the weight of a conformation are:

$$q_{hh} = e^{-\varepsilon_{hh}/kT} \qquad \text{for every hh contact}$$

$$s = s_0 e^{-\varepsilon_s/kT} \qquad \text{for every non-initial monomer in a helix}$$

$$\sigma s \qquad \text{for the first monomer in a helix}$$

6.2.2 Intersecting the Grammars

For the two-helix bundle problem, our grammar is the CFG (5.2), and our finite-state automaton is shown in Figure 6.3. It is more complicated than the Zimm-Bragg model because it tries to model the shape of a helix in a square lattice. Like the Zimm-Bragg model, it does not explicitly generate self-contacts; but it can be viewed as a cover grammar for a grammar which does (see Section 5.3.1).

The state q_c is for coil; the states q_{ij} are for helices, where i cycles through four values corresponding to the periodicity of the shape of a helix in a square lattice (see Figure 5.9a), and j is used to ensure that all helices are at least six monomers long.

The procedure for intersecting a context-free grammar with a finite-state automaton is due to Bar-Hillel et. al [10, pp. 149–150]. Given a CFG $G = \langle V, \Sigma, S, P \rangle$ and a finite-state automaton (without ε-transitions) $M = \langle \Sigma, Q, Q_0, Q_f, \delta \rangle$, the new CFG G' has nonterminal alphabet $Q \times V \times Q \cup S'$, where S' is a new start symbol not in V; and its production set consists of all productions of the form

$$\langle q_0, X, q_n \rangle \to \langle q_0, \alpha_1, q_1 \rangle \langle q_1, \alpha_2, q_2 \rangle \cdots \langle q_{n-1}, \alpha_n, q_n \rangle$$

where $X \to \alpha_1 \cdots \alpha_n \in P$ $(n > 0)$, and for each i, either $\alpha_i \in V$ or else $\alpha_i \in \Sigma$ and $\langle q_{i-1}, \alpha_i, q_i \rangle \in \delta$; and

$$\langle q, X, q \rangle \to \varepsilon$$

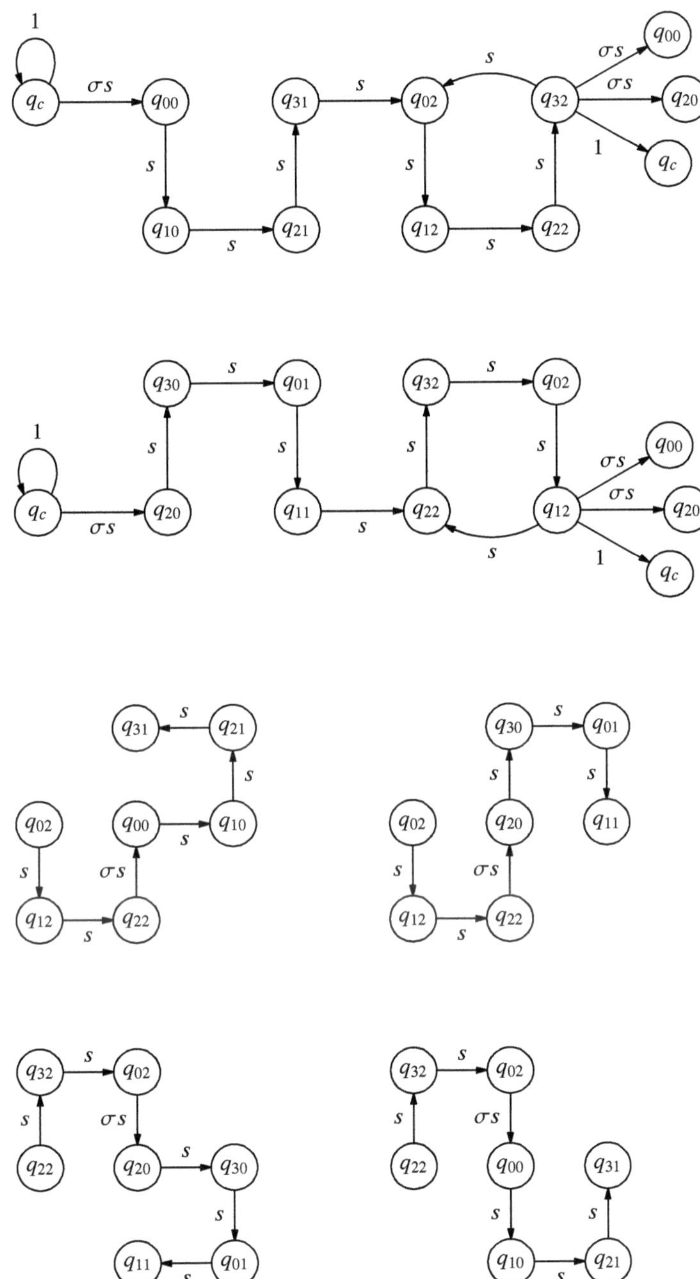

Fig. 6.3 Automaton for helices in a square lattice. Nodes with the same label represent the same state; the state is shown in multiple locations for visual clarity. The full automaton has the union of the transitions shown in all six diagrams. The initial states are $\{q_c, q_{00}, q_{20}\}$; the final states are $\{q_c, q_{12}, q_{32}\}$.

for all $q \in Q$, where $X \to \varepsilon \in P$; and, finally,

$$S' \to \langle q_0, S, q_f \rangle$$

for all $q_0 \in Q_0$, $q_f \in Q_f$. The resulting grammar generates the language $L(G) \cap L(M)$.

The difference between this construction and one like Brown and Wilson's for pseudoknots is that the two component grammars are fully integrated, so that we may let them control each other however we please. For example, when two helices come together to form a bundle, self-contacts should only be allowed between monomers on the sides of the helices facing each other. Our original CFG had the rule $X \to Z$ which generated a self-contact; in the intersected grammar its corresponding productions include those of the form $\langle q_{ij}, X, q_{i'j'} \rangle \to \langle q_{ij}, Z, q_{i'j'} \rangle$. We may now stipulate that $i, i' \in \{0, 2\}$ for such productions, which ensures that only one side of a helix may participate in nonlocal contacts.

6.2.3 Computing the Partition Function

We compute the partition function offline as described in Section 5.4, with some modifications. First, we incorporate the weights σ and s from the Zimm-Bragg model, and an additional factor q_{hh} for every helical hh contact.

Second, previously we estimated the number of conformations of a loop of length l as $\frac{1}{4}U(l)$ and the number of conformations of the tails with combined length l as $C(l-1)$. Now that these loops and tails may include helices, which are rigid, we must adjust these estimates. Our current approach is simply to count each helix as a single step in a neighbor-avoiding walk, without trying to take into account the length of the helix.

Finally, since the grammar is most error-prone with closed conformations, we use a special set of rules for loops of length eight or less, shown below (weights are conformation counts only):

$$X \xrightarrow{1} HHHX'HHH$$

$$X \xrightarrow{1} UX'HHH$$

$$X \xrightarrow{1} HHHX'U$$

$$X \xrightarrow{1} HXH$$

$$X \xrightarrow{1} X'X'X' \tag{6.1}$$

$$X \xrightarrow{4} CCCCCCC$$

$$X' \to X \mid C$$

$$U \to H \mid C$$

$$H \to \langle q,a,q' \rangle \qquad \langle q,a,q' \rangle \in \delta_H$$

$$C \to \langle q,a,q' \rangle \qquad \langle q,a,q' \rangle \in \delta_C$$

These rules do not exhaustively cover all possible loops of length eight or less; a number of possibilities were left out somewhat arbitrarily to limit overcounting. Chen and Dill's steric compatibility matrices might be a more principled solution.

6.2.4 Evaluation Against Exact Enumeration

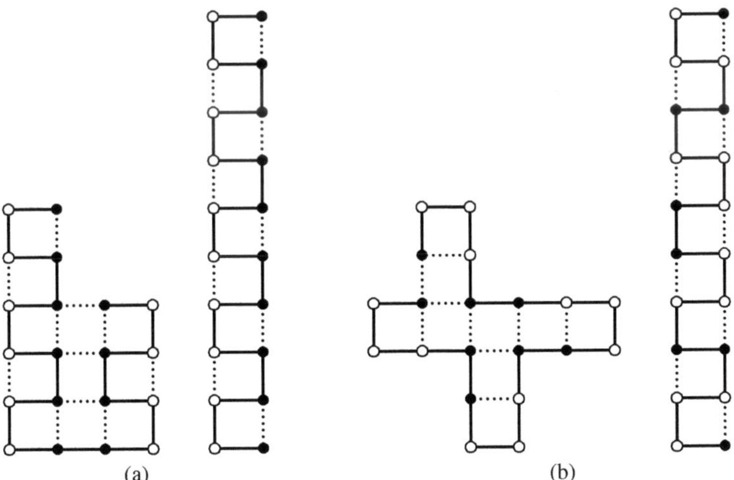

Fig. 6.4 Some favorable structures for example sequences: (a) hpphhpphhpphhpphhpph; (b) hppphhppphhppphhpphhpppph. Hydrophobic monomers (h) are indicated by black circles; polar monomers (p) by white circles.

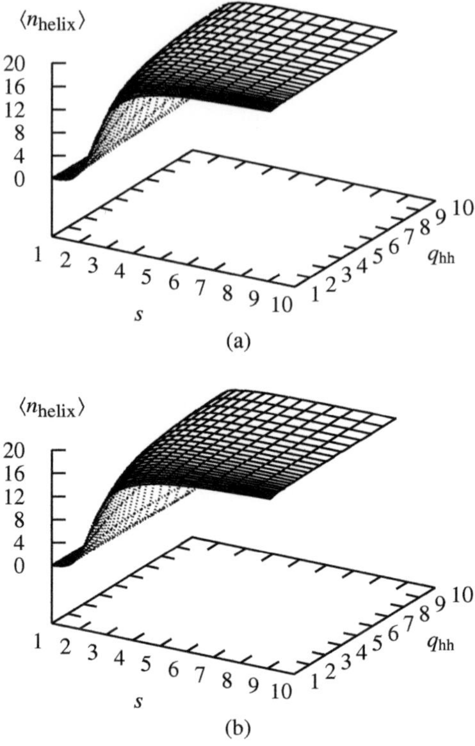

Fig. 6.5 Comparison against exact enumeration. Sequence: hpphhpphhpphhpphhpph; helix units versus s and q_{hh}. (a) Exact enumeration. (b) Parser.

We compared our parser against exact enumeration in various ways. First, we tried the sequence hpphhpphhpphhpphhpph, which has minimum-energy structures high in both helix units and hh contacts (Figure 6.4a). In this experiment and those below, $\sigma = 0.01$. Figure 6.5 shows the average number of helix units as a function of the parameters s and q_{hh}, and Figure 6.6 shows superimposed cross-sections of these functions; the output of the parser qualitatively agrees with that of the exact enumerator. Figures 6.7 and 6.8 compare the average number of hh contacts; again there is qualitative agreement, except for the region $s < 2$ of Figure 6.8b, which is not very important because such low values for s make helix units unfavorable relative to random coil, which is unrealistic.

We next tried the sequence hppphhppphhppphhppph, which has structures with many hh contacts and structures with many helix units, but not both at the same time (Figure 6.4b). Figures 6.9 and 6.10 compare the average number of helix units; as with the first sequence, the parser's output qualitatively agrees with the exact enumerator's.

Figures 6.11 and 6.12 compare the average number of hh contacts. The agreement is not as good as before for high q_{hh} due to both overcounting of some struc-

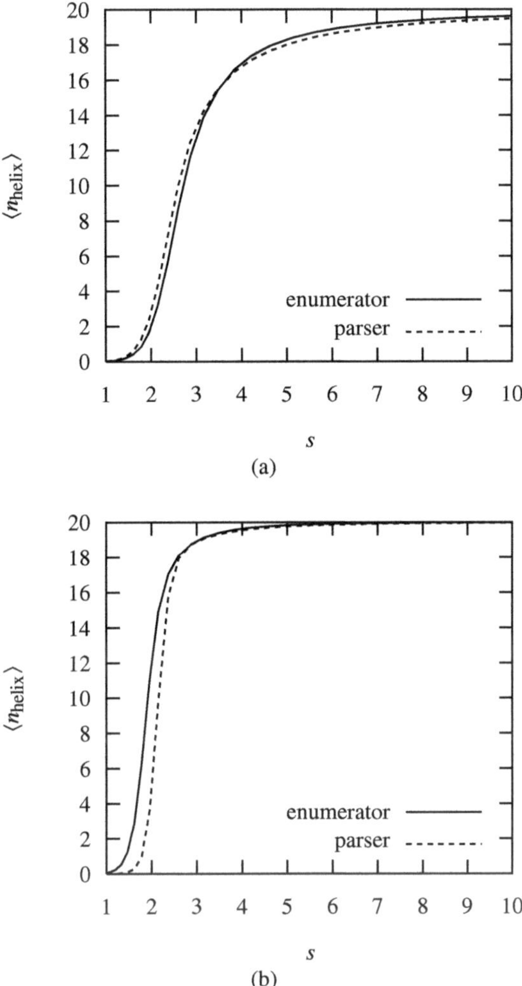

Fig. 6.6 Comparison against exact enumeration. Sequence: hpphhpphhpphhpphhpph; helix units versus s. (a) $q_{hh} = 1$. (b) $q_{hh} = 10$.

tures and undercounting of others. For example, the grammar is able to generate small spirals, but does not have the "memory" needed to keep the spiral from colliding with itself. Figure 6.13a shows an unviable conformation generated by the grammar. Such structures are causing the parser to overestimate the average number of hh contacts for high q_{hh} and low s (again, such low values are unrealistic). On the other hand, the grammar does not have a rule that would let helices contact each other at right angles. Figure 6.13a shows a viable conformation with three helices that the grammar does not generate. Missing this structure is causing the parser to underestimate the average number of hh contacts for high q_{hh} and s. Nevertheless,

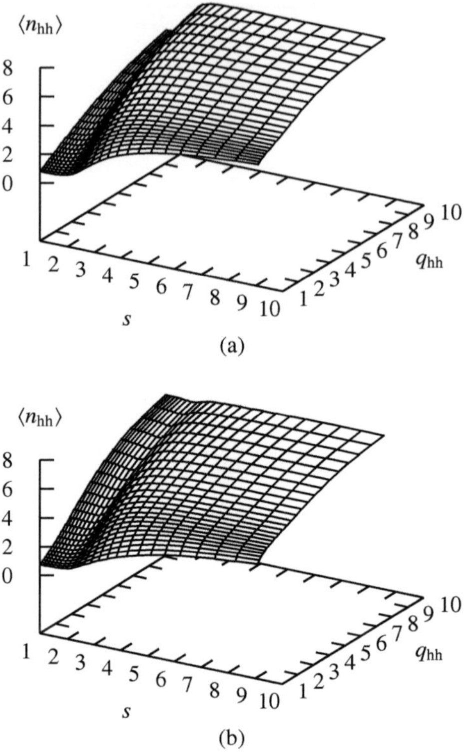

(a)

(b)

Fig. 6.7 Comparison against exact enumeration. Sequence: hpphhpphhpphhpphhpph; hh contacts versus s and q_{hh}. (a) Exact enumeration. (b) Parser.

the parser agrees with the enumerator in predicting that the number of hh contacts should decrease with s, in contrast to the first sequence.

The grammar could be modified to try to improve agreement, and this deserves further work. It is possible that in a three-dimensional lattice, the problem of collisions will be less severe because a greater proportion of structures will have viable conformations.

6.3 Intersection in Nonlinear sLMGs

In a simple LMG there are no restrictions on what literals may be conjoined in the right-hand side of a production. This makes sLMG closed under intersection: if S_1 and S_2 are the start symbols of two sLMGs G_1 and G_2 (with disjoint nonterminal alphabets), create a new start symbol S and add the production

$$S(x) :- S_1(x), S_2(x)$$

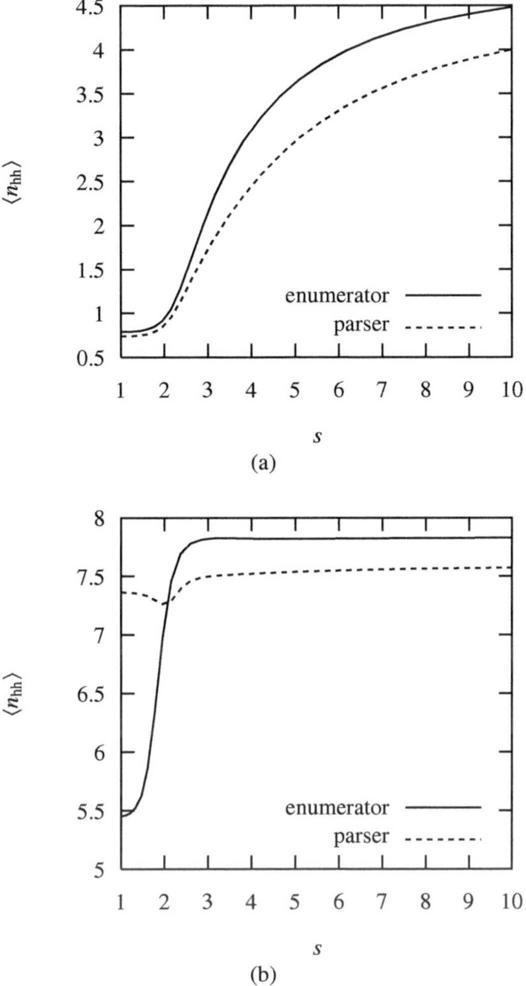

Fig. 6.8 Comparison against exact enumeration. Sequence: hpphhpphhpphhpphhpph; hh contacts versus s. (a) $q_{hh} = 1$. (b) $q_{hh} = 10$.

which recognizes $L(G_1) \cap L(G_2)$. Thus sLMG internalizes the intersection operation, which allows more control than Brown and Wilson's scheme. The caveats from our critique of that scheme still apply, however. For example, Boullier [19] gives a range concatenation grammar (which has an equivalent sLMG) which he claims models German scrambling, a construction in which all the nouns of a sentence can appear in any order. His grammar checks for a verb for every noun and vice versa, using intersection to enforce all these constraints simultaneously. But like Brown and Wilson's system, it relies on false assumptions about the generated string to ensure that the constraints are properly coordinated.

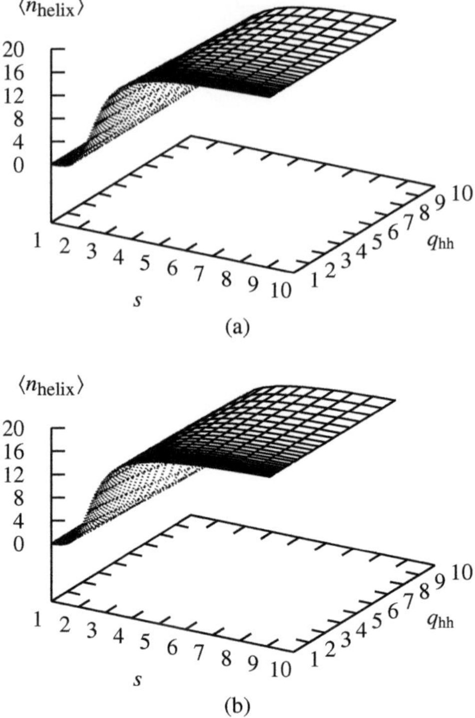

Fig. 6.9 Comparison against exact enumeration. Sequence: hppphhppphhppphhppph; helix units versus s and q_{hh}. (a) Exact enumeration. (b) Parser.

Nevertheless, nonlinear sLMG's closure under intersection might be a useful property for modeling complex folds like protein β-sheets. We start with some building blocks:

$$\text{Anti}(a_1 X, Y a_2) :- \text{Anti}(X, Y) \qquad\qquad a_i \in \Sigma$$

$$\text{Anti}(\varepsilon, \varepsilon)$$

$$\text{Par}(a_1 X, a_2 Y) :- \text{Par}(X, Y) \qquad\qquad a_i \in \Sigma$$

$$\text{Par}(\varepsilon, \varepsilon)$$

$$\text{Adj}(X, Y) :- \text{Ant}(X, Y)$$

$$\text{Adj}(X, Y) :- \text{Par}(X, Y)$$

The predicates Anti and Par generate pairs of adjacent antiparallel and parallel strands, respectively, and the predicate Adj generates two adjacent strands in either configuration. Irregularities as in Figure 5.16a are also possible, but not shown here.

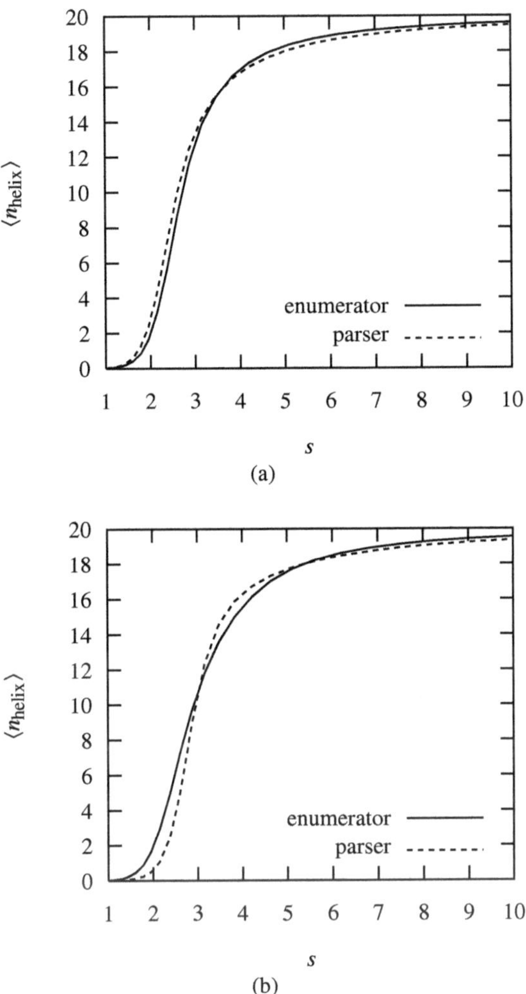

Fig. 6.10 Comparison against exact enumeration. Sequence: hppphhppphhppphhpppph; helix units versus s. (a) $q_{hh} = 1$. (b) $q_{hh} = 10$.

We can then use the intersection ability of sLMG to combine these pairs of strands into a sheet. Thus the following grammar generates β-sheets where the strands are arranged according to their order in the sequence:

$$\mathrm{Beta}(AB) :- \mathrm{B}(A,B)$$
$$\mathrm{B}(ABY,B') :- \mathrm{B}(A,B), \mathrm{Adj}(B,B')$$
$$\mathrm{B}(BY,B') :- \mathrm{Adj}(B,B')$$

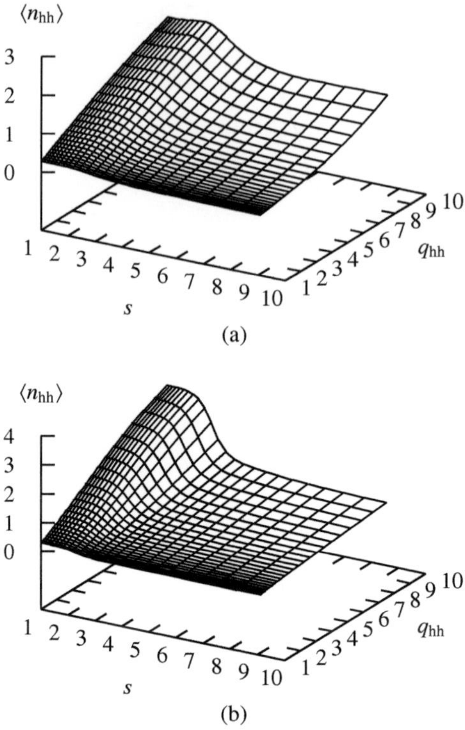

Fig. 6.11 Comparison against exact enumeration. Sequence: hppphhppphhppphhppph; hh contacts versus s and q_{hh}. (a) Exact enumeration. (b) Parser.

The first argument to B is a β-sheet minus the last strand, and the second argument is the last strand. The second production forms a larger β-sheet out of a smaller one by appending a new last strand and joining it to the previous last strand using Adj. This production has $\mathscr{O}(n^5)$ possible instantiations (because it takes six indices to specify the variables on the left-hand side, but the arguments of B are always adjacent, eliminating one index), and therefore the parsing complexity of this grammar is also $\mathscr{O}(n^5)$. Crucially, this complexity bound is not dependent on the number of strands, because each series of contacts is generated in sister subderivations, unlike the multicomponent TAG analysis.

But even sister subderivations can control each other via their root nonterminal (predicate) symbols, as illustrated in the following example. A β-sheet can be rolled into a cylinder to form a β-barrel (Figure 6.14). We can generate these as well, but we must keep track of the direction of each strand so as not to generate any Möbius strips, as in the grammar of Figure 6.15. Here B has three arguments: the first strand, the middle part, and the last strand; there is an additional predicate symbol B′ which is the same as B, except that B′ is for sheets with antiparallel first and last strands, whereas B is restricted here to sheets with parallel first and last strands. The first

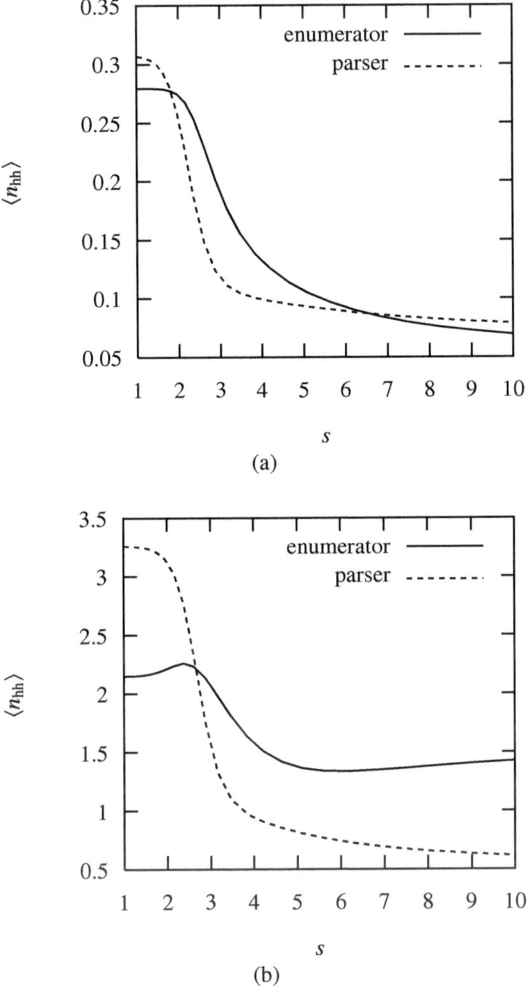

Fig. 6.12 Comparison against exact enumeration. Sequence: hppphhppphhppphhpppph; hh contacts versus s. (a) $q_{hh} = 1$. (b) $q_{hh} = 10$.

production joins the first and last strands to form a barrel; it uses the information in the B vs. B$'$ distinction to join the strands so that no Möbius strips will be generated.

The strands of β-sheets do not always appear in linear order; they can be permuted as in Figure 6.16. We can model such permutations by increasing the degree of synchronous parallelism (that is, the number of arguments to B), and therefore increasing parsing complexity. By contrast, since multicomponent TAG already uses synchronous parallelism to generate all the strands together, it allows permutations of strands at no extra cost.

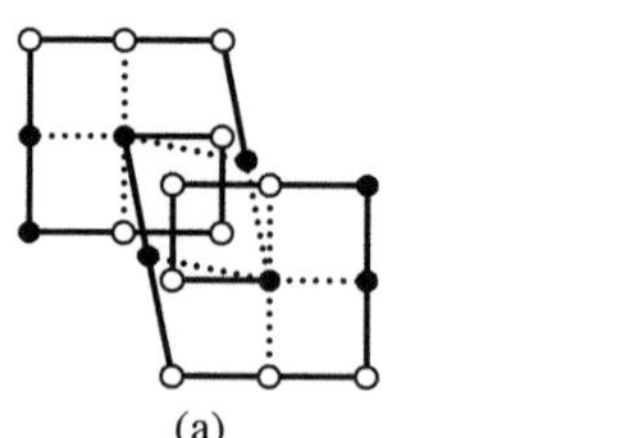 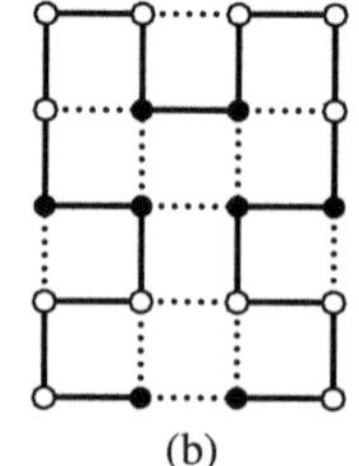

(a)   (b)

Fig. 6.13 Examples of grammar overcounting (a) and undercounting (b). Hydrophobic monomers (h) are indicated by black circles; polar monomers (p) by white circles.

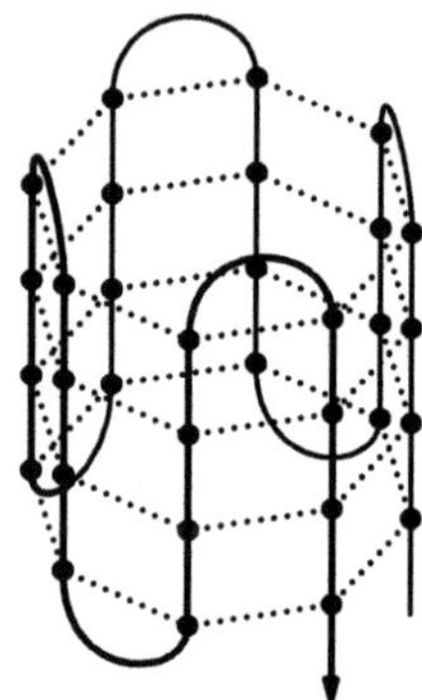

Fig. 6.14 β-barrel.

$$\mathrm{Barrel}(ABC) :- \mathrm{B}(A,B,C), \mathrm{Par}(A,C)$$
$$\mathrm{Barrel}(ABC) :- \mathrm{B}'(A,B,C), \mathrm{Anti}(A,C)$$
$$\mathrm{B}(A,BCY,C') :- \mathrm{B}'(A,B,C), \mathrm{Anti}(C,C')$$
$$\mathrm{B}(A,BCY,C') :- \mathrm{B}(A,B,C), \mathrm{Par}(C,C')$$
$$\mathrm{B}(A,Y,A') :- \mathrm{Par}(A,A')$$
$$\mathrm{B}'(A,BCY,C') :- \mathrm{B}(A,B,C), \mathrm{Anti}(C,C')$$
$$\mathrm{B}'(A,BCY,C') :- \mathrm{B}'(A,B,C), \mathrm{Par}(C,C')$$
$$\mathrm{B}'(A,Y,A') :- \mathrm{Anti}(A,A')$$

Fig. 6.15 sLMG for β-barrels.

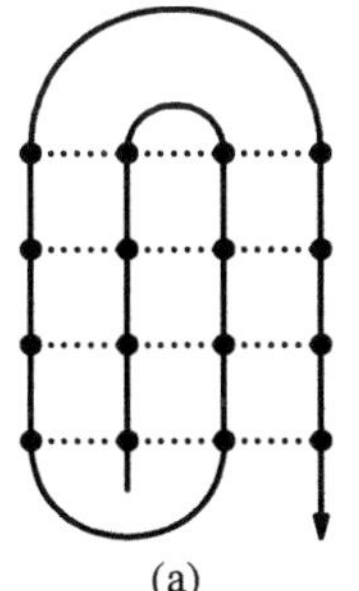 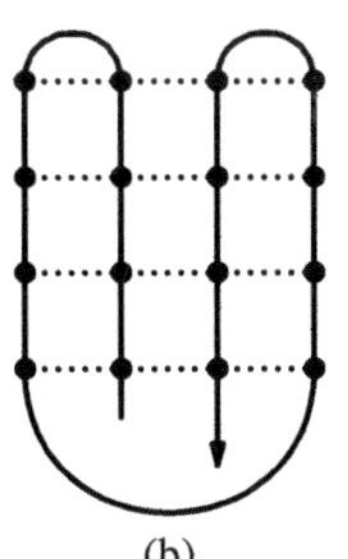 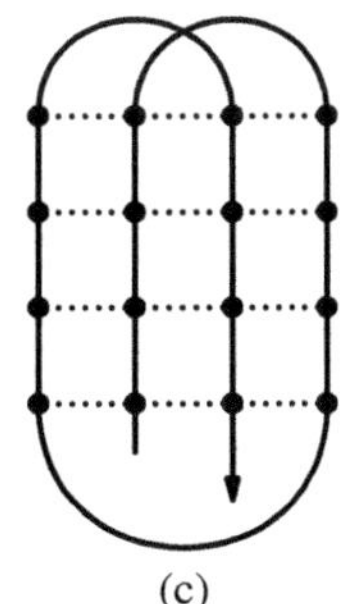

(a)   (b)   (c)

Fig. 6.16 Permuted β-sheets.

Suppose we envision a sheet being built up one strand at a time, each successive strand being added to either side of the sheet:

$$\mathrm{Beta}(ABCD) :- \mathrm{B}(A,B,C,D)$$
$$\mathrm{B}(ABC,D,Y,B') :- \mathrm{B}(A,B,C,D), \mathrm{Adj}(B,B')$$
$$\mathrm{B}(A,B,CDY,B') :- \mathrm{B}(A,B,C,D), \mathrm{Adj}(D,B')$$
$$\mathrm{B}(\varepsilon,B,Y,B') :- \mathrm{Adj}(B,B')$$

Figure 6.16a shows an example sheet that can be generated by this grammar but not the previous ones. In this grammar, the second and fourth arguments to B are the leftmost and rightmost strands (*not* respectively) in the folded structure. The second production adds a new strand on one side, and the third production adds a new strand on the other. Both productions have $\mathscr{O}(n^7)$ possible instantiations if we take into account that the four arguments to B will always be adjacent.

Suppose we always build up a sheet out of two smaller sheets:

$$\mathrm{Beta}(ABCDE) :- \mathrm{B}(A,B,C,D,E)$$
$$\mathrm{B}(ABC,D,EYA',B',C'D'E') :- \mathrm{B}(A,B,C,D,E), \mathrm{B}(A',B',C',D',E'), \mathrm{Adj}(B,D')$$
$$\mathrm{B}(A,B,CDEYA',B',C',D',E') :- \mathrm{B}(A,B,C,D,E), \mathrm{B}(A',B',C',D',E'), \mathrm{Adj}(D,D')$$
$$\mathrm{B}(\varepsilon,B,C,D,\varepsilon) :- \mathrm{Adj}(B,D)$$

Figure 6.16b shows an example sheet that can be generated by this grammar but not the previous ones. In this grammar, the second and fourth arguments are again the leftmost and rightmost strands (*not* respectively) in the folded structure. The second and third productions join two β-sheets together in two different ways; there are conceivably four ways to join them together, but using only these two avoids spurious ambiguity. Both productions have $\mathscr{O}(n^{12})$ possible instantiations if we take into account that the five arguments to B will always be adjacent.

Figure 6.16c shows the only permutation of four strands that the above grammar cannot generate. This does not seem problematic, since, at least for sheets formed out of two hairpin motifs, this permutation was not known as of 1991 to occur in nature [21, p. 31].

It should be emphasized, however, that any energies or conformation counts added to these grammars will not be able to make the self-contacts between two strands dependent on self-contacts with other strands. Akutsu [6] and Lyngsø and Pedersen [84] have shown that certain formulations of the problem of predicting RNA secondary structures with generalized pseudoknots are NP-hard. It turns out that both of these proofs assume some kind of dependence between nonadjacent strands. Akutsu assumes that no base can participate in two pairs (one on either side), which is true of RNA secondary structures but not of protein structures. Lyngsø and Pedersen assume that the energy of a base pairing (i, j) can be affected by another base pairing $(j - 1, i')$ even if i and i' are in different strands (or by $(j', i + 1)$ even if j and j' are in different strands); it remains to be seen whether such dependencies might be needed, for example, in calculating conformation counts for β-sheets.

6.4 Summary

Intersection is a technique which has been somewhat neglected in computational linguistics, but we have shown in this chapter that it has the potential to provide extra SGC in a way that is useful for analyzing biological sequences. There is a danger, however, of thinking that the extra WGC gained by intersection corresponds with extra SGC. We have explored several ways of employing this technique: we demonstrated a flaw in Brown and Wilson's use of intersections of CFLs for RNA pseudoknots, and we proposed intersections of CFGs and finite-state automata for two-helix bundles and nonlinear sLMGs for larger helix bundles and β-sheets. For future work we would like to extend the helix-bundle work to a three-dimensional lattice and bundles of three or more helices, using a TAG or a nonlinear sLMG. Since these larger bundles can be synthesized in the laboratory (Ken Dill, p.c.), such a model could be evaluated more directly.

Chapter 7
Conclusion

We began this study with the question: What makes one grammar formalism better than another? We developed a theoretical framework for dealing with this question, drawing on ideas from Miller and Joshi and others: in this framework we measure the generative power of grammars by choosing an interpretation domain suited to the task and restricting the interpretation functions to be defined on local domains. We applied this framework in three general areas: statistical parsing, machine translation, and biological sequence analysis.

Statistical parsing

The ability of a grammar formalism to describe statistical parsing models is characterized by its SGC with respect to the domain of weighted trees (or other weighted structures). This interpretation domain classifies formalisms rather coarsely: of the weakly context-free formalisms we examined, only tree-insertion grammar had greater statistical-modeling power than CFG. But this negative result led to a reinterpretation of lexicalized PCFGs as being cover grammars for grammars resembling TAG. This reinterpretation gave some new insights into how lexicalized PCFG parsers work and how to train them.

We demonstrated a model based on probabilistic tree-insertion grammar with sister-adjunction, discussing implementation details and some of its conceptual advantages. We described how to train this model both using heuristic reconstruction of structural descriptions and using EM to train directly on incomplete structural descriptions. Results on the Penn (English) Treebank were comparable to lexicalized PCFG parsers, and results on the Penn Chinese Treebank were state-of-the-art.

Machine translation

The ability of a grammar formalism to describe translations between languages is characterized by its SGC with respect to the domain of string (or tree) pairs. By contrast with the previous case, this interpretation domain classifies formalisms rather finely. Much progress has been made in applying synchronous grammars (CFG and TSG) to statistical machine translation, but a particularly challenging area has been tree-to-tree translation. Richer synchronous grammar formalisms may provide the power needed to model syntactic divergences, but because there are many different levels of power, the right formalism must be chosen. It would be incorrect to conclude from the failure of one formalism that other more powerful formalisms are not worth trying.

We focused on one synchronous formalism, synchronous regular-form TAG, as a basis for translation systems. We formally demonstrated that it has greater translation power than other synchronous formalisms currently used for statistical machine translation and provided some examples of how its extra power could be useful for difficult constructions.

Biological sequence analysis

The ability of a grammar formalism to describe secondary/tertiary structures of chain molecules is characterized by its SGC with respect to the domain of linked strings. This interpretation domain, like the previous one, classifies formalisms rather finely. We presented a synthesis of previous research in this area, discussing both CFG and formalisms beyond CFG. We showed how Chen and Dill's statistical-mechanical model can be recast as a weighted CFG, and how several kinds of secondary/tertiary structures can be modeled by variants of TAG.

We then turned our attention to the mechanism of intersection, which has a more natural interpretation in this domain than with syntactic structures. We took a close look at uses and abuses of this mechanism, and described two legitimate uses. First, we showed how to extend our CFG implementation of Chen and Dill's statistical-mechanical model to use a CFG intersected with a finite-state automaton, allowing it to model helix bundles, which was not previously feasible. Second, we discussed how nonlinear sLMGs might use intersection to model β-sheets more efficiently than previous approaches.

Conclusion

Grammars are gaining or regaining attention in various quarters of research in natural language processing and structural biology. But the theory fueling these applications will not be fully effective unless it is able to ask and answer the right questions: Is this grammar formalism more powerful than that grammar formalism for this particular application? We have set up a framework for carrying out such comparisons, and used it to explore three areas of application. We hope that the exploration will continue: empirical evaluation of theoretical results presented here, new results about other grammar formalisms, and investigation into further areas of application.

References

1. Abe, N., Mamitsuka, H.: Predicting protein secondary structures based on stochastic tree grammars. Machine Learning **29**, 275–301 (1997)
2. Abeille, A.: Parsing French with Tree Adjoining Grammar: some linguistic accounts. In: Proceedings of the Twelfth International Conference on Computational Linguistics (COLING), pp. 7–12 (1988)
3. Abney, S.P.: Stochastic attribute-value grammars. Computational Linguistics **23**, 597–618 (1997)
4. Aho, A.V., Ullman, J.D.: Syntax directed translations and the pushdown assembler. Journal of Computer and System Sciences **3**, 37–56 (1969)
5. Aho, A.V., Ullman, J.D.: Translations on a context free grammar. Information and Control **19**, 439–475 (1971)
6. Akutsu, T.: Dynamic programming algorithms for RNA secondary structure prediction with pseudoknots. Discrete Applied Mathematics **104**, 45–62 (2000)
7. Alberts, B., Johnson, A., Lewis, J., Raff, M., Roberts, K., Walter, P.: Molecular Biology of the Cell, 4th edn. Garland Science, New York (2002)
8. Ambati, V., Lavie, A.: Improving syntax driven translation models by re-structuring divergent and non-isomorphic parse tree structures. In: Proceedings of AMTA-2008 Student Research Workshop, pp. 235–244 (2008)
9. Baker, J.K.: Trainable grammars for speech recognition. In: Proceedings of the Spring Conference of the Acoustical Society of America, pp. 547–550 (1979)
10. Bar-Hillel, Y., Perles, M., Shamir, E.: On formal properties of simple phrase structure grammars. In: Y. Bar-Hillel (ed.) Language and Information: Selected Essays on their Theory and Application, pp. 116–150. Addison-Wesley, Reading, MA (1964)
11. Becker, T., Joshi, A., Rambow, O.: Long distance scrambling and tree adjoining grammar. In: Proceedings of the Fifth Conference of the European Chapter of the Association for Computational Linguistics (EACL), pp. 21–26 (1991)
12. Bertsch, E., Nederhof, M.J.: On the complexity of some extensions of RCG parsing. In: Proceedings of the Seventh International Workshop on Parsing Technologies (IWPT), pp. 66–77 (2001)
13. Bikel, D.M.: Intricacies of Collins' parsing model. Computational Linguistics **30**, 479–511 (2004)
14. Bikel, D.M.: On the parameter space of generative lexicalized statistical parsing models. Ph.D. thesis, University of Pennsylvania (2004)
15. Bikel, D.M., Chiang, D.: Two statistical parsing models applied to the Chinese Treebank. In: Proceedings of the Second Chinese Language Processing Workshop, pp. 1–6 (2000)
16. Bikel, D.M., Miller, S., Schwartz, R., Weischedel, R.: Nymble: a high-performance learning name-finder. In: Proceedings of the Fifth Conference on Applied Natural Language Processing (ANLP), pp. 194–201 (1997)

17. Bod, R.: A computational model of language performance: Data Oriented Parsing. In: Proceedings of the Fourteenth International Conference on Computational Linguistics (COLING), pp. 855–859 (1992)
18. Booth, T., Thompson, R.: Applying probability meausres to abstract languages. IEEE Transactions on Computers **C-22**(5), 442–450 (1973)
19. Boullier, P.: Chinese numbers, MIX, scrambling, and range concatenation grammars. In: Proceedings of the Ninth Conference of the European chapter of the Association for Computational Linguistics (EACL), pp. 53–60 (1999)
20. Boullier, P.: Range concatenation grammars. In: Proceedings of the Sixth International Workshop on Parsing Technologies (IWPT), pp. 53–64 (2000)
21. Branden, C.I., Tooze, J.: Introduction to Protein Structure, 2nd edn. Garland Publishing, Inc., New York (1999)
22. Bresnan, J., Kaplan, R.M., Peters, P.S., Zaenen, A.: Cross-serial dependencies in Dutch. Linguistic Inquiry **13**, 613–635 (1982)
23. Brown, M., Wilson, C.: RNA pseudoknot modeling using intersections of stochastic context free grammars with applications to database search. In: Proceedings of the Pacific Symposium on Biocomputing, pp. 109–125 (1996)
24. Brown, P.F., Della Pietra, S.A., Della Pietra, V.J., Mercer, R.L.: The mathematics of statistical machine translation: Parameter estimation. Computational Linguistics **19**, 263–311 (1993)
25. Charniak, E.: Statistical parsing with a context-free grammar and word statistics. In: Proceedings of the Fourteenth National Conference on Artificial Intelligence (AAAI), pp. 598–603. AAAI Press/MIT Press (1997)
26. Charniak, E.: A maximum-entropy-inspired parser. In: Proceedings of the First Meeting of the North American Chapter of the Association for Computational Linguistics (NAACL), pp. 132–139 (2000)
27. Chen, J., Vijay-Shanker, K.: Automated extraction of TAGs from the Penn Treebank. In: Proceedings of the Sixth International Workshop on Parsing Technologies (IWPT), pp. 65–76 (2000)
28. Chen, S.J., Dill, K.A.: Statistical thermodynamics of double-stranded polymer molecules. J. Chem. Phys. **103**(13), 5802–5813 (1995)
29. Chen, S.J., Dill, K.A.: Theory for the conformational changes of double-stranded chain molecules. J. Chem. Phys. **109**(11), 4602–4616 (1998)
30. Chen, S.J., Dill, K.A.: RNA folding energy landscapes. Proceedings of the National Academy of Sciences **97**(2), 646–651 (2000)
31. Chiang, D.: Statistical parsing with an automatically-extracted tree adjoining grammar. In: Proceedings of the 38th Annual Meeting of the Association for Computational Linguistics, pp. 456–463 (2000)
32. Chiang, D.: Constraints on strong generative power. In: Proceedings of the 39th Annual Meeting of the Association for Computational Linguistics, pp. 124–131 (2001)
33. Chiang, D.: Putting some weakly context-free formalisms in order. In: Proceedings of the Sixth International Workshop on TAG and Related Formalisms (TAG+), pp. 11–18 (2002)
34. Chiang, D.: Mildly context sensitive grammars for estimating maximum entropy parsing models. In: Proceedings of Formal Grammar 2003. CSLI Publications (2003)
35. Chiang, D.: A hierarchical phrase-based model for statistical machine translation. In: Proceedings of the 43rd Annual Meeting of the Association for Computational Linguistics, pp. 263–270 (2005)
36. Chiang, D.: Hierarchical phrase-based translation. Computational Linguistics **33**(2), 201–228 (2007)
37. Chiang, D., Joshi, A.K.: Formal grammars for estimating partition functions of double-stranded chain molecules. In: Proceedings of the Second International Conference on Human Language Technology Research (HLT), pp. 63–67 (2002)
38. Chiang, D., Scheffler, T.: Flexible composition and delayed tree-locality. In: Proceedings of the Ninth International Workshop on TAG and Related Formalisms (TAG+), pp. 17–24 (2008)

39. Chiang, D., Schuler, W., Dras, M.: Some remarks on an extension of synchronous TAG. In: Proceedings of the Fifth International Workshop on TAG and Related Formalisms (TAG+), pp. 61–66 (2000)
40. Chiou, F.D., Chiang, D., Palmer, M.: Facilitating treebank annotation using a statistical parser. In: Proceedings of the First International Conference on Human Language Technology Research (HLT), pp. 117–120 (2001). Poster session
41. Chomsky, N.: Formal properties of grammars. In: R.D. Luce, R.R. Bush, E. Galanter (eds.) Handbook of Mathematical Psychology, vol. 2, pp. 323–418. Wiley, New York (1963)
42. Chomsky, N., Miller, G.A.: Introduction to the formal analysis of natural languages. In: R.D. Luce, R.R. Bush, E. Galanter (eds.) Handbook of Mathematical Psychology, vol. 2, pp. 269–321. Wiley, New York (1963)
43. Clark, S., Curran, J.R.: Log-linear models for wide-coverage CCG parsing. In: Proceedings of the 2003 Conference on Empirical Methods in Natural Language Processing (EMNLP), pp. 97–104 (2003)
44. Collins, M.: Three generative lexicalised models for statistical parsing. In: Proceedings of ACL-EACL, pp. 16–23 (1997)
45. Collins, M.: Head-driven statistical models for natural language parsing. Ph.D. thesis, University of Pennsylvania (1999)
46. Collins, M., Brooks, J.: Prepositional phrase attachment through a backed-off model. In: Proceedings of the Third Workshop on Very Large Corpora (WLVC), pp. 27–38 (1995)
47. Dempster, A.P., Laird, N.M., Rubin, D.B.: Maximum likelihood from incomplete data via the EM algorithm. Journal of the Royal Statistical Society, Series B **39**, 1–38 (1977)
48. DeNeefe, S., Knight, K.: Synchronous tree adjoining machine translation. In: Proceedings of the 2009 Conference on Empirical Methods in Natural Language Processing (EMNLP), pp. 727–736 (2009)
49. Dras, M.: A meta-level grammar: redefining synchronous TAG for translation and paraphrase. In: Proceedings of the 37th Annual Meeting of the Association for Computational Linguistics, pp. 80–87 (1999)
50. Dras, M., Bleam, T.: How problematic are clitics for S-TAG translations? In: Proceedings of the Fifth International Workshop on TAG and Related Formalisms (TAG+), pp. 241–244 (2000). Poster session
51. Eisner, J.: Learning non-isomorphic tree mappings for machine translation. In: Companion Volume to the Proceedings of the 41st Annual Meeting of the Association for Computational Linguistics, pp. 205–208 (2003)
52. Galley, M., Graehl, J., Knight, K., Marcu, D., DeNeefe, S., Wang, W., Thayer, I.: Scalable inference and training of context-rich syntactic translation models. In: Proceedings of COLING-ACL 2006, pp. 961–968 (2006)
53. Galley, M., Hopkins, M., Knight, K., Marcu, D.: What's in a translation rule? In: Proceedings of HLT-NAACL 2004, pp. 273–280 (2004)
54. Gildea, D.: Corpus variation and parser performance. In: Proceedings of the 2001 Conference on Empirical Methods in Natural Language Processing (EMNLP), pp. 167–202 (*sic*) (2001)
55. Gildea, D.: Loosely tree-based alignment for machine translation. In: Proceedings of the 41st Annual Meeting of the Association for Computational Linguistics, pp. 80–87 (2003)
56. Goodman, J.: Global thresholding and multiple-pass parsing. In: Proceedings of the Second Conference on Empirical Methods in Natural Language Processing (EMNLP), pp. 11–25 (1997)
57. Groenink, A.V.: Surface without structure: Word order and tractability issues in natural language analysis. Ph.D. thesis, University of Utrecht (1997)
58. Han, C.H., Potter, D., Storoshenko, D.R.: Non-local right-node raising: an analysis using delayed tree-local MC-TAG. In: Proceedings of the Tenth International Workshop on TAG and Related Formalisms (TAG+) (2010)
59. Hopcroft, J.E., Ullman, J.D.: Introduction to Automata Theory, Languages, and Computation. Addison-Wesley, Reading, MA (1979)
60. Huang, L., Knight, K., Joshi, A.: Statistical syntax-directed translation with extended domain of locality. In: Proceedings of AMTA 2006, pp. 65–73 (2006)

61. Huybregts, M.A.C.: Overlapping dependencies in Dutch. Utrecht Working Papers in Linguistics **1**, 24–65 (1976)
62. Huybregts, R.: The weak inadequacy of context-free phrase structure grammars. In: G. de Haan, M. Trommele, W. Zonneveld (eds.) Van Periferie naar Kern, pp. 81–99. Foris, Dordrecht (1984)
63. Hwa, R.: An empirical evaluation of probabilistic lexicalized tree insertion grammars. In: Proceedings of COLING-ACL, pp. 557–563 (1998)
64. Hwa, R.: Learning probabilistic lexicalized grammars for natural language processing. Ph.D. thesis, Harvard University (2001)
65. Johnson, M.: PCFG models of linguistic tree representations. Computational Linguistics **24**, 613–632 (1998)
66. Johnson, M., Geman, S., Canon, S., Chi, Z., Riezler, S.: Estimators for stochastic "unification-based" grammars. In: Proceedings of the 37th Annual Meeting of the Association for Computational Linguistics, pp. 535–541 (1999)
67. Joshi, A.K.: Tree adjoining grammars: How much context-sensitivity is necessary for assigning structural descriptions? In: D. Dowty, L. Karttunen, A. Zwicky (eds.) Natural Language Parsing, pp. 206–250. Cambridge University Press, Cambridge (1985)
68. Joshi, A.K.: A note on the strong and weak generative powers of formal systems. Theoretical Computer Science **293**, 243–259 (2003). Originally presented at TAG+5 in 2000
69. Joshi, A.K., Levy, L., Takahashi, M.: Tree adjunct grammars. Journal of Computer and System Sciences **10**, 136–163 (1975)
70. Joshi, A.K., Schabes, Y.: Tree-adjoining grammars. In: G. Rosenberg, A. Salomaa (eds.) Handbook of Formal Languages and Automata, vol. 3, pp. 69–124. Springer-Verlag, Heidelberg (1997)
71. Kato, Y., Seki, H., Kasami, T.: Subclasses of tree adjoining grammar for RNA secondary structure. In: Proceedings of the Seventh International Workshop on TAG and Related Formalisms (TAG+), pp. 48–55 (2004)
72. Klein, D., Manning, C.D.: Accurate unlexicalized parsing. In: Proceedings of the 41st Annual Meeting of the Association for Computational Linguistics, pp. 423–430 (2003)
73. Knuth, D.E.: Semantics of context-free languages. Mathematical Systems Theory **2**(2), 127–145 (1968). Corrections published in 5(1):95–96
74. Koehn, P., Och, F.J., Marcu, D.: Statistical phrase-based translation. In: Proceedings of HLT-NAACL 2003, pp. 127–133 (2003)
75. Kroch, A.S., Santorini, B.: The derived structure of the West Germanic verb raising construction. In: R. Freidin (ed.) Principles and Parameters in Comparative Grammar, pp. 269–338. MIT Press, Cambridge, MA (1991)
76. Kuroda, S.Y.: A topological study of phrase-structure languages. Information and Control **30**, 307–379 (1976)
77. Kuroda, S.Y.: A topological approach to structural equivalence of formal languages. In: A. Manaster-Ramer (ed.) Mathematics of Language, pp. 173–189. John Benjamins (1987)
78. Lambek, J.: The mathematics of sentence structure. American Mathematical Monthly **65**(3), 154–170 (1958)
79. Lari, K., Young, S.J.: The estimation of stochastic context-free grammars using the inside-outside algorithm. Computer Speech and Language **4**, 35–56 (1990)
80. Lau, K.F., Dill, K.A.: A lattice statistical mechanics model of the conformation and sequence spaces of proteins. Macromolecules **22**, 3986–3997 (1989)
81. Lewis, P.M., Stearns, R.E.: Syntax-directed transduction. Journal of the ACM **15**, 465–488 (1968)
82. Liu, Y., Liu, Q., Lin, S.: Tree-to-string alignment template for statistical machine translation. In: Proceedings of COLING-ACL 2006, pp. 609–616 (2006)
83. Liu, Y., Lü, Y., Liu, Q.: Improving tree-to-tree translation with packed forests. In: Proceedings of ACL-IJCNLP 2009, pp. 558–566 (2009)
84. Lyngsø, R.B., Pedersen, C.N.S.: RNA pseudoknot prediction in energy-based models. Journal of Computational Biology **7**(3/4), 409–427 (2000)

85. Madras, N., Slade, G.: The Self-Avoiding Walk. Birkhäuser, Boston (1993)
86. Magerman, D.M.: Statistical decision-tree models for parsing. In: Proceedings of the 33rd Annual Meeting of the Association for Computational Linguistics, pp. 276–283 (1995)
87. Marcus, M.P., Santorini, B., Marcinkiewicz, M.A.: Building a large annotated corpus of English: the Penn Treebank. Computational Linguistics **19**, 313–330 (1993)
88. Matsuzaki, T., Miyao, Y., Tsujii, J.: Probabilistic CFG with latent annotations. In: Proceedings of the 43rd Annual Meeting of the Association for Computational Linguistics, pp. 75–82 (2005)
89. Miller, P.H.: Strong Generative Capacity: The Semantics of Linguistic Formalism. CSLI Publications, Stanford (1999)
90. Miller, S., Fox, H., Ramshaw, L., Weischedel, R.: A novel use of statistical parsing to extract information from text. In: Proceedings of the First Meeting of the North American Chapter of the Association for Computational Linguistics (NAACL), pp. 226–233 (2000)
91. Miyao, Y., Ninomiya, T., Tsujii, J.: Probabilistic modeling of argument structures including non-local dependencies. In: Proceedings of the Conference on Recent Advances in Natural Language Processing (RANLP-2003), pp. 285–291 (2003)
92. Nesson, R., Shieber, S.M., Rush, A.: Induction of probabilistic synchronous tree-insertion grammars for machine translation. In: Proceedings of AMTA 2006 (2006)
93. Neumann, G.: Automatic extraction of stochastic lexicalized tree grammars from treebanks. In: Proceedings of the Fourth International Workshop on TAG and Related Formalisms (TAG+), pp. 120–123 (1998)
94. Nijholt, A.: Context-Free Grammars: Covers, Normal Forms, and Parsing. Springer-Verlag, Berlin (1980)
95. Och, F.J., Ney, H.: Improved statistical alignment models. In: Proceedings of the 38th Annual Meeting of the Association for Computational Linguistics, pp. 440–447 (2000)
96. Pentus, M.: Lambek grammars are context-free. In: Proceedings of the 8th Annual IEEE Symposium on Logic in Computer Science, pp. 429–433. Los Alamitos, CA (1993)
97. Pentus, M.: Lambek calculus is NP-complete. Tech. Rep. TR-2003005, CUNY Ph.D. Program in Computer Science (2003)
98. Petrov, S., Barrett, L., Thibaux, R., Klein, D.: Learning accurate, compact, and interpretable tree annotation. In: Proceedings of COLING-ACL 2006, pp. 433–440 (2006)
99. Post, E.L.: Formal reductions of the general combinatorial decision problem. American Journal of Mathematics **65**, 197–215 (1943)
100. Prescher, D.: Head-driven PCFGs with latent-head statistics. In: Proceedings of the Ninth International Workshop on Parsing Technologies (IWPT), pp. 115–124 (2005)
101. Pullum, G.K., Gazdar, G.: Natural languages and context-free languages. Linguistics and Philosophy **4**, 471–504 (1982)
102. Rambow, O., Satta, G.: Independent parallelism in finite copying parallel rewriting systems. Theoretical Computer Science **223**, 87–120 (1999)
103. Rambow, O., Vijay-Shanker, K., Weir, D.: D-tree grammars. In: Proceedings of the 33rd Annual Meeting of the Association for Computational Linguistics, pp. 151–158 (1995)
104. Ratnaparkhi, A.: A maximum-entropy model for part-of-speech tagging. In: Proceedings of EMNLP, pp. 133–142 (1996)
105. Resnik, P.: Probabilistic tree-adjoining grammar as a framework for statistical natural language processing. In: Proceedings of the Fourteenth International Conference on Computational Linguistics (COLING), pp. 418–424 (1992)
106. Rivas, E., Eddy, S.R.: The language of RNA: a formal grammar that includes pseudoknots. Bioinformatics **16**(4), 334–340 (2000)
107. Rogers, J.: Capturing CFLs with tree adjoining grammars. In: Proceedings of the 32nd Annual Meeting of the Association for Computational Linguistics, pp. 155–162 (1994)
108. Rogers, J.: Syntactic structures as multi-dimensional trees. Research on Language and Computation **1**, 265–305 (2003)
109. Sakakibara, Y., Brown, M., Hughey, R., Mian, I., Sjölander, K., Underwood, R.C., Haussler, D.: Stochastic context-free grammars for tRNA modeling. Nucleic Acids Research **22**, 5112–5120 (1994)

110. Satta, G., Schuler, W.: Restrictions on tree adjoining languages. In: Proceedings of COLING-ACL, pp. 1176–1182 (1998)

111. Schabes, Y.: Mathematical and computational aspects of lexicalized grammars. Ph.D. thesis, University of Pennsylvania (1990). Available as technical report MS-CIS-90-48

112. Schabes, Y.: Stochastic lexicalized tree-adjoining grammars. In: Proceedings of the Fourteenth International Conference on Computational Linguistics (COLING), pp. 426–432 (1992)

113. Schabes, Y., Abeille, A., Joshi, A.K.: Parsing strategies with 'lexicalized' grammars: Application to Tree Adjoining Grammars. In: Proceedings of the Twelfth International Conference on Computational Linguistics (COLING), pp. 578–583 (1988)

114. Schabes, Y., Shieber, S.M.: An alternative conception of tree-adjoining derivation. Computational Linguistics **20**, 91–124 (1994)

115. Schabes, Y., Waters, R.C.: Lexicalized context-free grammars. In: Proceedings of the 31st Annual Meeting of the Association for Computational Linguistics, pp. 121–129 (1993)

116. Schabes, Y., Waters, R.C.: Tree insertion grammar: a cubic-time parsable formalism that lexicalizes context-free grammar without changing the trees produced. Computational Linguistics **21**, 479–513 (1995)

117. Schabes, Y., Waters, R.C.: Stochastic lexicalized tree-insertion grammar. In: H. Bunt, M. Tomita (eds.) Recent Advances in Parsing Technology, pp. 281–294. Kluwer, Dordrecht (1996)

118. Schuler, W.: Preserving semantic dependencies in synchronous tree adjoining grammar. In: Proceedings of the 37th Annual Meeting of the Association for Computational Linguistics, pp. 88–95 (1999)

119. Schuler, W., Chiang, D., Dras, M.: Multi-component TAG and notions of formal power. In: Proceedings of the 38th Annual Meeting of the Association for Computational Linguistics, pp. 448–455 (2000)

120. Searls, D.B.: The linguistics of DNA. American Scientist **80**, 579–591 (1992)

121. Seki, H., Matsumura, T., Fujii, M., Kasami, T.: On multiple context-free grammars. Theoretical Computer Science **88**, 191–229 (1991)

122. Shieber, S.M.: Evidence against the context-freeness of natural language. Linguistics and Philosophy **8**(3), 333–344 (1985)

123. Shieber, S.M.: Restricting the weak generative capacity of synchronous tree-adjoining grammars. Computational Intelligence **10**(4), 371–385 (1994). Special Issue on Tree Adjoining Grammars

124. Shieber, S.M., Schabes, Y.: Synchronous tree-adjoining grammars. In: Proceedings of the Thirteenth International Conference on Computational Linguistics (COLING), vol. 3, pp. 1–6 (1990)

125. Shieber, S.M., Schabes, Y., Pereira, F.C.N.: Principles and implementation of deductive parsing. Journal of Logic Programming **24**, 3–36 (1995)

126. Storoshenko, D.R., Han, C.H.: Binding variables in english: An analysis using delayed tree locality. In: Proceedings of the Tenth International Workshop on TAG and Related Formalisms (TAG+) (2010)

127. Uemura, Y., Hasegawa, A., Kobayashi, S., Yokomori, T.: Tree adjoining grammars for RNA structure prediction. Theoretical Computer Science **210**, 277–303 (1999)

128. Vijay-Shanker, K., Weir, D.J.: The use of shared forests in tree adjoining grammar parsing. In: Proceedings of the Sixth Conference of the European Chapter of the Association for Computational Linguistics (EACL), pp. 384–393 (1993)

129. Weir, D.J.: Characterizing mildly context-sensitive grammar formalisms. Ph.D. thesis, University of Pennsylvania (1988)

130. Wu, D.: A polynomial-time algorithm for statistical machine translation. In: Proceedings of the 34th Annual Meeting of the Association for Computational Linguistics, pp. 152–158 (1996)

131. Wu, D.: Stochastic inversion transduction grammars and bilingual parsing of parallel corpora. Computational Linguistics **23**, 377–404 (1997)

132. Xia, F.: Extracting tree adjoining grammars from bracketed corpora. In: Proceedings of the 5th Natural Language Processing Pacific Rim Symposium (NLPRS-99), pp. 398–403 (1999)
133. Xia, F., Palmer, M., Xue, N., Okurowski, M.E., Kovarik, J., Chiou, F.D., Huang, S., Kroch, T., Marcus, M.: Developing guidelines and ensuring consistency for Chinese text annotation. In: Proceedings of the Second International Conference on Language Resources and Evaluation (LREC-2000), pp. 3–10. Athens, Greece (2000)
134. Yamada, K., Knight, K.: A syntax-based statistical translation model. In: Proceedings of the 39th Annual Meeting of the Association for Computational Linguistics, pp. 523–530 (2001)
135. Yamada, K., Knight, K.: A decoder for syntax-based statistical MT. In: Proceedings of the 40th Annual Meeting of the Association for Computational Linguistics, pp. 303–310 (2002)
136. Zhang, M., Jiang, H., Aw, A., Li, H., Tan, C.L., Li, S.: A tree sequence alignment-based tree-to-tree translation model. In: Proceedings of ACL-08: HLT, pp. 559–567 (2008)
137. Zimm, B.H., Bragg, J.K.: Theory of the phase transition between helix and random coil in polypeptide chains. Journal of Chemical Physics **31**, 526–535 (1959)

Keterangan

Index